Lecture Notes in Mathematics

2043

Editors:
J.-M. Morel, Cachan
B. Teissier, Paris

T0092388

For further volumes:
http://www.springer.com/series/304

Thomas H. Otway

The Dirichlet Problem for Elliptic-Hyperbolic Equations of Keldysh Type

 Springer

Thomas H. Otway
Department of Mathematical Sciences
Yeshiva University
New York
USA

ISBN 978-3-642-24414-8   e-ISBN 978-3-642-24415-5
DOI 10.1007/978-3-642-24415-5
Springer Heidelberg Dordrecht London New York

Lecture Notes in Mathematics ISSN print edition: 0075-8434
                              ISSN electronic edition: 1617-9692

Library of Congress Control Number: 2011943496

Mathematics Subject Classification (2010): 35-XX, 35M10

Printed on acid-free paper

Springer is part of Springer Science+Business Media (www.springer.com)

Preface

Partial differential equations of mixed elliptic–hyperbolic type arise in diverse areas of physics and geometry, including fluid and plasma dynamics, optics, cosmology, traffic engineering, projective geometry, geometric variational theory, and the theory of isometric embeddings. And yet even the linear theory of these equations is at a very early stage. This course examines various Dirichlet problems which can be formulated for equations of Keldysh type, one of the two main classes of linear elliptic–hyperbolic equations. Open boundary conditions (in which data are prescribed on only part of the boundary) and closed boundary conditions (in which data are prescribed on the entire boundary) are both considered. Emphasis is on the formulation of boundary conditions for which solutions can be shown to exist in an appropriate function space. Specific applications to plasma physics, optics, and analysis on projective spaces are discussed.

These notes were written to supplement a series of ten lectures given at Henan University in the summer of 2010. They are intended for graduate students and researchers in pure or applied analysis. In particular, the reader is expected to have a background in functional analysis – including Sobolev spaces – and the basic theory of partial differential equations, but not necessarily any prior expertise in the theory of mixed elliptic–hyperbolic equations. A familiarity with the geometry of differential forms is assumed in Sect. 5.6, but that material can be skipped without loss of continuity. Although some mathematical ideas which are used frequently in the text are collected in Chap. 2, that material is not intended to replace the above-listed prerequisites.

As is typical with monographs of this kind, some of the results have not previously appeared in the literature. Examples are Theorems 3.1–3.3, 6.2, and 6.3. But the proofs of those results are based on rather standard arguments for this field. We also include examples of results which have been asserted in the literature but for which detailed proofs are missing or obscure; an example of such "folklore" is Theorem 4.1. In all cases the form of the included results has been dictated more by expository considerations than by a desire to extend the research literature.

There has been considerable research into elliptic–hyperbolic equations in China, Russia, and Bulgaria. Unfortunately, some of the older literature of those countries remains difficult to access, and there are undoubtedly important contributions which I have missed. I would be grateful to the readers for comments on lacunae in the references and, of course, on errors or omissions in the text.

I am grateful to Daniela Lupo and Kevin R. Payne for discussion of Proposition 4.1, and to Antonella Marini for discussion related to Theorem 6.3. I am also grateful to Yisong Yang for suggesting the course on which these notes are based (and for many conversations about mathematical physics and analysis), to Yuxi Zheng for independently suggesting that I write a set of lecture notes on the subject (and also for many conversations about mathematical physics and analysis), and to Ke Wu for inviting me to give the lectures at Henan University. These notes have benefited substantially from the comments of anonymous referees for *Lecture Notes in Mathematics*. It has been a pleasure to work with the excellent editorial/production staff assigned to this series. Finally, I thank Shouxin Chen and the administration, faculty, and graduate students of the School of Mathematics and Information Science at Henan University for their hospitality during the week in which the lectures were given, and my fellow lecturer Joel Spruck for his collegiality during that time.

New York City   *Thomas H. Otway*

Contents

Chapter 1
Introduction

In the introductory chapters to most plasma physics texts (e.g., [43,45]), an idealized model is presented in which the plasma ion and electron temperatures – rather than being millions of degrees – are set to absolute zero. This is done to reduce the mathematics to its simplest possible form, but even in this case a rigorous description of the plasma is problematic: the classical Dirichlet problem, which is the physically natural boundary value problem in this context, turns out to be ill-posed on typical domains.

This kind of difficulty characterizes elliptic–hyperbolic equations that result from applying Maxwell's equations to various kinds of plasma waves, or to models for optical waves in the neighborhood of a caustic. Analogous difficulties arise in certain geometric variational problems that imply harmonic fields in projective space or on relativistically rotating discs. These notes are devoted to formulating and solving well-posed Dirichlet problems for elliptic–hyperbolic equations of *Keldysh type*, which is the class of partial differential equations that governs these examples.

1.1 What Is an Equation of Keldysh Type?

In the neighborhood of any point in their domain, second-order linear equations of elliptic, hyperbolic, or parabolic type can be put into the form of a potential, wave, or heat equation, respectively, by a nonsingular change of coordinates. (A brief review of the criteria for the type of a differential equation is given in Sect. 2.1.) There are analogous local canonical forms for equations that actually *change* from elliptic to hyperbolic type along a smooth curve; but in that case there is a greater variety of canonical local forms.

Consider a second-order, linear differential operator L acting on a scalar function $u(x, y)$, for $(x, y) \in \Omega \subset\subset \mathbf{R}^2$. Suppose that L is of elliptic type on a subdomain Ω^+, of hyperbolic type on a subdomain Ω^-, and of parabolic type on a smooth curve $\mathscr{G}$ (the *parabolic curve*) separating Ω^+ and Ω^-. Then it is always

T.H. Otway, *The Dirichlet Problem for Elliptic-Hyperbolic Equations of Keldysh Type*,
Lecture Notes in Mathematics 2043, DOI 10.1007/978-3-642-24415-5_1,
© Springer-Verlag Berlin Heidelberg 2012

possible to find a nonnegative integer m, and a real, nonsingular transformation of the independent variables, which in a neighborhood of some given point on $\mathcal{G}$ transforms L into one of two canonical forms, either

$$L_T \equiv y^{2m+1} u_{xx} + u_{yy} + \text{lower-order terms} \qquad (1.1)$$

or

$$L_K \equiv u_{xx} + y^{2m+1} u_{yy} + \text{lower-order terms}. \qquad (1.2)$$

A differential equation for which the differential operator L can be transformed locally into an operator of the form L_T is said to be of *Tricomi type*, whereas a differential equation for which the differential operator L can be transformed locally into an operator of the form L_K is said to be of *Keldysh type*.

These canonical forms were introduced by Cinquini-Cibrario in 1932 [8]. In the 1950s, emphasis shifted from the construction of special coordinates to the qualitative behavior of the coefficients of the higher-order terms. This emphasis, which was motivated by physical applications [13], resulted in the more general classes of canonical forms which we will construct explicitly in Sect. 3.2. Briefly, one begins by noticing that given a second-order linear differential operator L, it is always possible to introduce a nonsingular local transformation under which we obtain an operator having the form

$$Lu = \mathcal{K}(x, y) u_{xx} + u_{yy} + \text{lower-order terms}. \qquad (1.3)$$

One can, by a further nonsingular transformation, arrive at a coordinate system in which the *type-change function* $\mathcal{K}(x, y)$ has either of two forms:

1. $\mathcal{K}(y)$, with $\mathcal{K}(0) = 0$ and $y \mathcal{K}(y) > 0$ for $y \neq 0$, or
2. $\mathcal{K}(x)$, with $\mathcal{K}(0) = 0$ and $x \mathcal{K}(x) > 0$ for $x \neq 0$

The former characterizes equations of Tricomi type type and the latter, equations of Keldysh type.

Still more generally, one notices that the distinction between Tricomi type and Keldysh type is only meaningful in a neighborhood of the parabolic curve $\mathcal{G}$, and thus equations of Keldysh type can be characterized by the degeneration of their characteristic lines, which intersect $\mathcal{G}$ tangentially. This characterization, unlike the previous two, suggests that solutions to equations of Keldysh type might have relatively unpleasant analytic properties at the parabolic transition; and this tends to be the case.

1.1.1 Remarks

i) The precise relation between equations of Keldysh type, as described above, and the behavior of characteristic lines near the parabolic curve will be made explicit in Sect. 3.2.

ii) The word "type" in the context of L_T and L_K is not used in the same sense as it is used in defining operators of elliptic, hyperbolic, and parabolic type: the difference between Tricomi type and Keldysh type does not correspond to a difference in the sign of the discriminant associated to the operator L (or of its principal symbol), or in the existence of real roots of the associated quadratic form [20].

iii) The Tricomi/Keldysh distinction is only made here for equations in the plane, although it could, of course, be formulated more generally; see, e.g., [38]. We emphasize that, except for brief discussions in Sects. 6.4.2, 6.4.5, and Appendix B, this course will restrict itself to equations in two dimensions. We will further restrict our attention to equations which are linear to begin with or which can be linearized by a hodograph transformation. It is a measure of the difficulty of the topic that after all these severe restrictions, we can still make very few general assertions of the kind routinely made about classes of uniformly elliptic, parabolic, or hyperbolic equations.

iv) In addition to the restrictions listed in item *iii*), there appears to be a further restriction in that the boundary conditions considered are exclusively homogeneous. Unlike the restrictions of item *iii*), this one is more apparent than real. This is because the equations considered tend to be homogeneous but have acquired in this treatment an inhomogeneous term. The Dirichlet problem for an inhomogeneous equation having homogeneous boundary conditions is dual, in a sense that can be made precise (Sect. 2.6), to a Dirichlet problem for a homogeneous equation having inhomogeneous boundary conditions. Nevertheless, many of the techniques here do not have obvious extensions to inhomogeneous equations having inhomogeneous boundary conditions (Problem 17, Appendix B).

v) The context in which Keldysh himself worked was not elliptic–hyperbolic equations, but rather elliptic equations having a singularity resulting from degeneracy on the boundary [22]. There are many important results on elliptic or hyperbolic equations having a boundary singularity of Keldysh type (see, for examples of the former, the work of Čanić and Keyfitz [5] and for examples of the latter, the work of Jang and Masmoudi [21]). We exclude this topic, as our focus is not really on the work of Keldysh at all, but on mathematical and physical consequences of the earlier work of Cinquini-Cibrario.

1.2 Why Study Equations of Keldysh Type?

Any linear elliptic–hyperbolic partial differential equation of second order is locally of either Tricomi or Keldysh type; and the Dirichlet problem is typically ill-posed in the Tricomi case as well. But the important applications of Tricomi-type equations – to aerodynamics and to the isometric embedding of Riemannian manifolds – are already well represented in the expository literature. The vast mathematical literature on transonic aerodynamics which accumulated during the first half of the twentieth century is reviewed in [4], and the literature of the second half is reviewed in [6]; see also [41]. The relation of these applications to hyperbolic conservation laws and the theory of shock waves is also treated in both older [10] and more recent [11,42,47] expositions. An introduction to the isometric embedding of Riemannian manifolds via elliptic–hyperbolic equations is provided in [18]; for very recent approaches, see [7,17].

In a review of [41] written shortly before her death, Cinquini-Cibrario characterized that survey as "incomplete," as her own contributions published in the early 1930s had been ignored [9]. Those contributions had resulted in the introduction of equations of Keldysh type, and Cinquini-Cibrario's criticism could have been applied to the expository literature as a whole. For this reason we restrict our attention to precisely that class of elliptic–hyperbolic equations which has been neglected in the expository literature up to now.

Many of the techniques which we apply to equations of Keldysh type were originally applied to the somewhat more accessible equations of Tricomi type. The methods of Lupo, Morawetz and Payne [27, 28] in particular, which were introduced for equations of Tricomi type, underlie many of the arguments in Chaps. 3 and 4. Similarly, work by Barros-Neto and Gelfand [1–3] for the Tricomi equation apparently motivated the construction by S-X. Chen of a fundamental solution to the equation of Cinquini-Cibrario (Sect. 3.7).

Differential operators associated with equations of Tricomi type tend to be of real principal type [46], whereas those associated with equations of Keldysh type tend not to be. The major analytic properties for operators of real principal type depend only on the principal symbol. As a result, microlocal arguments work nicely for equations of Tricomi type; see [14, 15] and, especially, [38–40]. But microlocal arguments tend to fail when applied to equations of Keldysh type, the regularity of which depends delicately on the form of lower-order terms. Despite this major difference, the analytic methods for the two canonical forms are generally more alike than they are different. Thus, despite their decreased regularity, a study of equations of Keldysh type provides a reasonable introduction to the general topic of linear, second-order equations of mixed elliptic–hyperbolic type.

1.3 Objectives and Organization of the Course

There are two common objectives in writing an expository monograph on a field of active research. The first is to provide, for a field that has reached a certain level of maturity, a homogeneous treatment which organizes the creative chaos of the research literature into an orderly and self-consistent body of results. The second is to provide, for a field that has not yet matured, a strong argument for more intensive study and suggested directions for such study. In the context of elliptic equations, the first kind of monograph is illustrated by Morrey's review of variational theory [35] or the monograph of Ladyzhenskaya and Ural'tseva [25]; the second kind of monograph is illustrated by Stoker's review of water-wave theory [44].

Although any exposition should distinguish what is known from what is not known, in the second kind of monograph the current state of the theory may be hard to describe. This is because results in an emerging field tend to arise from the study of interesting special cases, and the effects of serendipitous aspects of such cases may be difficult to assess. For example, in models of zero-temperature plasma, the type-change function is of the special form $\mathcal{K}(x, y) = x - y^2$ (Chap. 4). In deriving analytic properties of that model, it is customary to exploit the natural symmetries of a parabola. Also, there tends in emerging fields to be a shortage of "sharp" results (those for which counter-examples exist that show that a result cannot be improved in a given direction).

Thus in Chap. 3 we give a "toy model" for what a generic theory might look like, emphasizing the simplest possible type-change function of Keldysh type: the function $\mathcal{K}(x) = x$. An obvious direction of research would be to replace the hypothesis that the type-change function has this explicit form by one of the three characterizations of equations of Keldysh type given in Sect. 1.2. We show in Sect. 3.3 that there is "something to study": the classical Dirichlet problem is shown to be ill-posed on a typical domain. We then proceed to consider weaker hypotheses (*i.e.*, problems in which we specify the solution on only part of the boundary) or weaker results (solutions which may not be differentiable), and find in Sects. 3.4 and 3.5 that we can construct well-posed boundary value problems. On the one hand, many of the results clearly generalize beyond functions having the form $\mathcal{K}(x) = x$, and some of these generalizations are discussed in subsequent chapters. On the other hand, we argue in Sect. 3.6 that this particular type-change function has physical interest – in connection with atmospheric and space plasmas – as well as historical significance.

The discussions of Chap. 3 and later chapters rely on certain fundamental analytic methods, some of which are collected in Chap. 2. That review is actually a good deal shorter than it could be. This has been done in order to keep the amount of preparatory material as brief as possible, in order not to delay for too long the introduction of the core topics in Sects. 3.1–3.4. So for example, the technical discussions of D-star-shaped domains, the hodograph map, and Hodge–Bäcklund transformations are postponed until Sects. 3.5.1, 5.3, and 5.6, respectively. A chapter appendix, Sect. 2.7, is devoted to a partial review of applications.

Chapter 4 is devoted to one of the most significant applications of equations of Keldysh type. The example comes from models of zero-temperature plasma to which a magnetic field has been applied. As already noted, the classical Dirichlet problem is the natural one for stationary forms of Maxwell's equations in a Coulomb gauge, and yet is over-determined in this case. Thus in the only plasma model simple enough to be accessible to rigorous mathematical analysis, that analysis suggests that the model is ill-posed! And yet, techniques introduced by Lupo, Morawetz and Payne [27] for equations of Tricomi type can be extended to this model. Moreover, the mathematical resolution of the problem – that is, the construction of a weighted function space in which sufficiently weak solutions can be shown to exist uniquely – yields physical insight: evidence for a heating zone at any point for which the resonance frequency of the applied field is tangent to a flux surface [36].

In Chap. 5 we review another area in which equations of Keldysh type arise: non-geometrical optics. Optical analogies play an important role in the theory of partial differential equations, and in fact the methods introduced in this chapter illuminate applications introduced in preceding chapters. In particular, the use of Bäcklund transformations in a series of fundamental papers by Magnanini and Talenti [29–33] holds the promise of unifying, to some extent, the applications of elliptic–hyperbolic equations reviewed in Sect. 2.7. Thus in Sect. 5.6 we employ a method for relating mass densities used in diverse applications of elliptic–hyperbolic equations. Also in Chap. 5, we discuss the hodograph map. This is the technical link between the physical models underlying elliptic–hyperbolic theory, which are generally nonlinear, and the linear boundary value problems which tend to be the objects suitable for mathematical analysis.

A relatively recent development in the study of equations of Tricomi type is their increasingly important role in the problem of isometric embeddings: the question of the extent to which one can "visualize" a Riemannian surface by sitting it in three-dimensional Euclidean space [7, 18, 19, 23, 26]. While equations of Keldysh type do not seem to arise naturally in this geometric context, they do arise naturally in two other areas of geometric analysis: waves in space-time and on projective spaces. This material is introduced in Chap. 6. The Laplace–Beltrami equation on the extended projective disc is of Keldysh type. The Laplace–Beltrami equation on a two-dimensional reduction of a space-time metric may be of either Tricomi or Keldysh type, depending on the original metric and on the reduction. (It may also be of elliptic or hyperbolic type on the entire domain.) However, in an important special case it is of Keldysh type. This is the case of a metric for a stationary, rotating, axisymmetric field in which the time and axial coordinates have been neglected. The Laplace–Beltrami operator on such metrics includes the Laplace–Beltrami operator on the extended projective disc as a special case. Closed, time-like curves, which are associated with causality violation, are permitted on the hyperbolic region of the equation. (This is in distinction to a broadly analogous model for equations of Tricomi type; see the discussion of (3.22) in [40] and also [37] and the references therein.) A model for a relativistically rotating disc on which closed, time-like curves are permitted is one of several models discussed in the appendices to these notes (Appendix B, Problem 11).

In addition to open problems and suggested directions for research in the future, the appendices include a review of the various notions of solution employed in the text, a collection of the main results, and a rough comparison of the state of the art for the two classes of equations. Khuri's method for the use of anisotropic function spaces in closed elliptic–hyperbolic boundary value problems [24] was announced too recently to influence the literature (and thus this text); but it is discussed briefly in Appendix A.6, and also in item *ii*) of Problem 4 in Appendix B.

1.4   The Problem of Smoothness

A problem in any text on analysis is how much discussion should be devoted to the optimal smoothness of coefficients, domain boundaries, etc. (The problem is exacerbated somewhat by the fact that some authors use the term "smooth" to mean

C^1 and others use it to mean C^∞. Unless otherwise stated, we use the term in its latter meaning.)

From one point of view, these are technical side issues, which distract the attention of the reader from the main ideas of the course. Prior to the twentieth century, it was common to impose the (generally unstated) assumption that the data are smooth enough for the operations indicated in a proof to be performed – which is sometimes a short step from the assumption that the data are smooth enough for the assertions of the theorem to be true. This heuristic approach still can be found in parts of the physics and engineering literature.

From a different point of view, if questions of smoothness are side issues, then most of modern analysis is a side issue. In particular, the subject of this text is how the smoothness requirements in the classical Dirichlet problem can be relaxed in order to permit singular solutions to problems which lack C^2 solutions; so if we ignore the question of smoothness, we can ignore the problems addressed in the text.

Clearly some smoothness questions are more critical than others. Consider the following two examples:

The smoothness of weight functions, introduced in Sect. 2.4.2 and used frequently thereafter, is significant. Because the weighted versions of the function space H_0^1 are defined as completions, the integral expressions may only make sense as limits of a sequence of regular approximations, and the existence of a weak gradient cannot be assumed; see, e.g., [12]. So if the weight functions are not sufficiently smooth, then the various definitions of weak solution listed in Sect. A.1 may not be equivalent. In this text we will assume – unless stated otherwise – that weight functions have bounded derivative and do not vanish on a set of positive measure.

On the other hand, it is appropriate that the coefficients b and c in Sect. 2.4.3 are described only as "sufficiently smooth." In the expected applications, one or the other of the coefficients may have a jump in one of its derivatives along the parabolic curve, and the nature of that possible discontinuity will depend on the details of the specific application.

In any assertion, we use the term "sufficiently smooth" to mean that specifying the precise (or optimal) degree of smoothness is of secondary importance in that assertion.

1.5 The Problem of Lower-Order Terms

Equations of Keldysh type are notorious for the dependence of their analytic properties on the precise form of the lower-order terms. This property is not characteristic of equations of Tricomi type, as noted in Sect. 1.2.

1.5.1 First-Order Terms

Consider as an example the equation

$$\mathscr{K}(x, y) u_{xx} + k \mathscr{K}_x(x, y) u_x + u_{yy} = 0, \qquad (1.4)$$

where k is a constant. If $k = 1$, the analytic properties of solutions appear to be unusually strong. Indeed, for many choices of $\mathscr{K}$ that arise in important applications, the only satisfying results are for the case $k = 1$; see, e.g., Theorems 4.2 or 5.2. If on the other hand $k = 1/2$, the analytic properties of solutions to (1.4) appear to be especially weak. For example, corresponding to the two existence theorems cited for the case $k = 1$ are two nonexistence theorems for the case $k = 1/2$ – Theorem 3.1 (c.f. [34]) and Theorem 6.1; see also Theorem 6.2.

There are unusually strong results for $k = 1$ because for that value of k the differential operator is formally self-adjoint, which allows powerful functional-analytic methods to be applied. The case $k = 1/2$ is problematic because for that value of k the differential operator satisfies a conservation law, which can be used to show that boundary value problems are well-posed when data are prescribed on only part of the domain boundary (Sect. 3.3); boundary value problems with data prescribed on the full boundary become over-determined. In that sense, the case $k = 1$ is just symmetric enough to be useful, whereas the case $k = 1/2$ is too symmetric to be useful. Note that the higher-order terms for equations of Tricomi type are always formally self-adjoint, so this is an important difference between the two canonical forms.

1.5.2 Zeroth-Order Terms

Regardless of type, the very existence of a zeroth-order term will generally prevent the linearization of a quasilinear equation by the hodograph method (Sect. 5.3), or the analysis of the equation by first-order methods such as those of Sect. 2.5. But there are more direct examples of the influence of zeroth-order terms on the existence of solutions to equations of Keldysh type. Gu ([16], Theorem 3; Sect. 6.4.2, below) introduced examples of an equation of Keldysh type in $\mathbf{R}^n$ which will not have solutions if the coefficient of the zeroth-order term is not of sufficiently large magnitude (although the quantity involved is also a coefficient of the first-order term). The solution to the closed boundary value problem of Theorem 4.3 will not exist unless the coefficient of the zeroth-order term has sufficiently large magnitude (although there is an additional, *smallness* condition on the coefficient of the first-order term). Terms of order zero are only occasionally considered in this text.

1.6   Notation and Conventions

Unless otherwise indicated, a subscripted variable denotes (usually partial) differentiation in the direction of the variable, whereas subscripted numbers denote components of a matrix, vector, or tensor. We also occasionally use the shorthand notation $\partial_i \equiv \partial/\partial x^i$. Differentiation of vector or matrix components in the direction of a variable is indicated by a comma preceding the subscripted variable. The default coordinate system is cartesian, in which the subscript 1 denotes a component projected onto the x-axis, the subscript 2 denotes a component projected onto the y-axis, and the subscript 3 denotes a component projected onto the z-axis. In particular, $\mathbf{x} = (x_1, x_2, x_3) = (x, y, z)$ and we denote the canonical cartesian basis by $\left(\hat{i}, \hat{j}, \hat{k} \right)$. Repeated indices are summed over the number of dimensions. In proving inequalities, we frequently denote by C generic positive constants, the value of which may change from line to line.

Borrowing the terminology of fluid dynamics, we will refer to the smooth curve $\mathscr{G}$ along which the type of the equation degenerates in a *parabolic transition*, as the *sonic curve*. The motivation for this terminology is the physical fact that the continuity equation of gas dynamics changes type at the speed of sound.

Unless otherwise indicated, every domain Ω is bounded, connected and has piecewise C^1 boundary $\partial\Omega$ oriented in a counterclockwise direction. Occasionally we will emphasize one or more of these properties by repeating it in the statement of a result; but the properties are assumed to hold whether stated or not. For non-triviality, we assume that every domain in the text includes an arc of the sonic curve for the corresponding equation.

References

1. Barros-Neto, J., Gelfand, I.M.: Fundamental solutions for the Tricomi operator. Duke Math. J. **98**, 465–483 (1999)
2. Barros-Neto, J., Gelfand, I.M.: Fundamental solutions for the Tricomi operator II. Duke Math. J. **111**, 561–584 (2002)
3. Barros-Neto, J., Gelfand, I.M.: Fundamental solutions for the Tricomi operator III, Duke Math. J. **128**, 119–140 (2005)
4. Bers, L.: Mathematical Aspects of Subsonic and Transonic Gas Dynamics. Wiley, New York (1958)
5. Čanić, S., Keyfitz, B.: A smooth solution for a Keldysh type equation. Commun. Partial Diff. Equations **21**, 319–340 (1996)
6. Chapman, C.J.: High Speed Flow. Cambridge University Press, Cambridge (2000)
7. Chen, G-Q., Slemrod, M., Wang, D.: Isometric immersions and compensated compactness. Commun. Math. Phys. **294**, 411–437 (2010)
8. Cibrario, M.: Sulla riduzione a forma canonica delle equazioni lineari alle derivate parziali di secondo ordine di tipo misto. Rendiconti del R. Insituto Lombardo, Ser. II **65**, 889–906 (1932)
9. Cinquini-Cibrario, M.: Review of Rassias, John M. (GR-UATH) Lecture notes on mixed type partial differential equations, World Scientific Publishing Co., Teaneck, 1990. In: Mathematical Reviews, MR1082555 (91m:35162)

10. Courant, R., Friedrichs, K.O.: Supersonic Flow and Shock Waves. Interscience, New York (1948)
11. Dafermos, C.M.: Hyperbolic Conservation Laws in Continuum Mechanics. Springer, Berlin (2005)
12. Fabes, E., Kenig, C., Serapioni, R.: The local regularity of solutions of degenerate elliptic equations. Commun. Partial Diff. Equations **7**, 77–116 (1982)
13. Frankl', F.L.: Problems of Chaplygin for mixed sub- and supersonic flows [in Russian]. Izv. Akad. Nauk SSSR, ser. mat. **9**(2), 121–143 (1945)
14. Gramchev, T.V.: An application of the analytic microlocal analysis to a class of differential operators of mixed type. Math. Nachr. **121**, 41–51 (1985)
15. Groothuizen, R.J.P.: Mixed Elliptic-Hyperbolic Partial Differential Operators: A Case Study in Fourier Integral Operators. CWI Tract, vol. 16, Centrum voor Wiskunde en Informatica, Amsterdam (1985)
16. Gu, C.: On partial differential equations of mixed type in n independent variables. Commun. Pure Appl. Math. **34**, 333–345 (1981)
17. Han, Q.: Local solutions to a class of Monge-Ampère equations of mixed type. Duke Math. J. **136** 421–473 (2007)
18. Han, Q., Hong, J.-X.: Isometric Embedding of Riemannian Manifolds in Euclidean Spaces, Mathematical Surveys and Monographs, vol. 130, American Mathematical Society, Providence (2006)
19. Han, Q., Khuri, M.: On the local isometric embedding in $\mathbf{R}^3$ of surfaces with Gaussian curvature of mixed sign. Commun. Anal. Geom. **18**, 649–704 (2010)
20. Hersh, R.: How to classify differential polynomials. Amer. Math. Monthly **80**, 641–654 (1973)
21. Jang, J., Masmoudi, N.: Well-posedness for compressible Euler equations with physical vacuum singularity. Commun. Pure Appl. Math. **62**, 1327–1385 (2009)
22. Keldysh, M.V.: On certain classes of elliptic equations with singularity on the boundary of the domain [in Russian]. Dokl. Akad. Nauk SSSR **77**, 181–183 (1951)
23. Khuri, M.A.: The local isometric embedding in $\mathbf{R}^3$ of two-dimensional Riemannian manifolds with Gaussian curvature changing sign to finite order on a curve. J. Differential Geom. **76**, 249–291 (2007)
24. Khuri, M.A.: Boundary value problems for mixed type equations and applications. J. Nonlinear Anal. Ser. A: TMA **74**, 6405–6415 (2011)
25. Ladyzhenskaya, O.A., Ural'tseva, N.N.: Linear and Quasilinear Elliptic Equations. Academic Press, New York (1968)
26. Lin, C.S.: The local isometric embedding in $\mathbf{R}^3$ of a 2-dimensional Riemanian manifolds with Gaussian curvature changing sign cleanly. Commun. Pure Appl. Math. **39**, 867–887 (1986)
27. Lupo, D., Morawetz, C.S., Payne, K.R.: On closed boundary value problems for equations of mixed elliptic-hyperbolic type. Commun. Pure Appl. Math. **60**, 1319–1348 (2007)
28. Lupo, D., Morawetz, C.S., Payne, K.R.: Erratum: "On closed boundary value problems for equations of mixed elliptic-hyperbolic type," [Commun. Pure Appl. Math. **60**, 1319–1348 (2007)]. Commun. Pure Appl. Math. **61**, 594 (2008)
29. Magnanini, R., Talenti, G.: On complex-valued solutions to a 2D eikonal equation. Part one: qualitative properties. Contemp. Math. **283**, 203–229 (1999)
30. Magnanini, R., Talenti, G.: Approaching a partial differential equation of mixed elliptic-hyperbolic type. In: Anikonov, Yu.E., Bukhageim, A.L., Kabanikhin, S.I., Romanov, V.G. (eds.) Ill-posed and Inverse Problems, pp. 263–276. VSP, Utrecht (2002)
31. Magnanini, R., Talenti, G.: On complex-valued solutions to a two-dimensional eikonal equation. II. Existence theorems. SIAM J. Math. Anal. **34**, 805–835 (2003)
32. Magnanini, R., Talenti, G.: On complex-valued solutions to a 2D eikonal equation. III. Analysis of a Bäcklund transformation. Appl. Anal. **85**, 249–276 (2006)
33. Magnanini, R., Talenti, G.: On complex-valued 2D eikonals. IV. Continuation past a caustic. Milan J. Math. **77**, 1–66 (2009)
34. Morawetz, C.S., Stevens, D.C., Weitzner, H.: A numerical experiment on a second-order partial differential equation of mixed type. Commun. Pure Appl. Math. **44**, 1091–1106 (1991)

35. Morrey, C.B.: Multiple Integrals in the Calculus of Variations. Springer, Berlin (1966)
36. Otway, T.H.: Unique solutions to boundary value problems in the cold plasma model. SIAM J. Math. Anal. **42**, 3045–3053 (2010)
37. Parker, P.E.: Geometry of bicharacteristics. In: Advances in Differential Geometry and General Relativity, Contemp. Math., vol. 359, pp. 31–40. American Mathematical Society, Providence (2004)
38. Payne, K.R.: Propagation of singularities for solutions to the Dirichlet problem for equations of Tricomi type, Rend. Sem. Mat. Univ. Pol. Torino **54**, 115–137 (1996)
39. Payne, K.R.: Interior regularity of the Dirichlet problem for the Tricomi equation. J. Mat. Anal. Appl. **199**, 271–292 (1996)
40. Payne, K.R.: Solvability theorems for linear equations of Tricomi type. J. Mat. Anal. Appl. **215**, 262–273 (1997)
41. Rassias, J.M.: Lecture notes on Mixed Type Partial Differential Equations. World Scientific, Teaneck (1990)
42. Smoller, J.: Shock Waves and Reaction-Diffusion Equations. Springer, Berlin (1983)
43. Stix, T.H.: The Theory of Plasma Waves. McGraw-Hill, New York (1962)
44. Stoker, J.J.: Water Waves. Interscience, New York (1987)
45. Swanson, D.G.: Plasma Waves. Institute of Physics, Bristol (2003)
46. Taylor, M.E.: Pseudodifferential Operators. Princeton University Press, Princeton (1981)
47. Zheng, Y.: Systems of Conservation Laws: Two-Dimensional Riemann Problems. Birkhauser, Boston (2001)

Chapter 2
Mathematical Preliminaries

The purpose of this chapter is to emphasize some standard material in functional analysis and the theory of partial differential equations which will be particularly useful in subsequent chapters. In addition, a brief survey of applications is given in Sect. 2.7. Specialists in partial differential equations may prefer to skip Sects. 2.1–2.6 of this chapter.

2.1 Boundary Value Problems for Elliptic–Hyperbolic Equations

The specification of type for a linear partial differential equation in the plane largely determines the expected boundary value problems for the equation. If, for example, the equation is of elliptic type, then in two space dimensions we expect the Dirichlet problem to be well-posed on a smooth domain that is topologically a disc. Similarly, we expect the Cauchy problem to be well-posed for a hyperbolic equation in a cone (and in the reflection of that cone about the spatial plane), and an initial boundary value problem with Dirichlet conditions at the base to be well-posed for a parabolic equation in a cylinder (Fig. 2.1).

No analogous expectation exists for linear partial differential equations which change from elliptic to hyperbolic type across a smooth curve in the plane. Rather, the literature on such equations presents a disorderly collection of special cases. For many years, elliptic–hyperbolic equations were nearly exclusively associated with transonic fluid flow, and physical reasoning determined the kinds of boundary value problems that were posed. However, such equations are now ubiquitous in physics and geometry, and one expects a well-posed boundary value problem to be derivable from purely mathematical reasoning.

We restrict our attention to linear second-order partial differential equations on a domain $\Omega \subset \mathbf{R}^2$. Write the differential operator L acting on the second-order terms in the form

T.H. Otway, *The Dirichlet Problem for Elliptic-Hyperbolic Equations of Keldysh Type*,
Lecture Notes in Mathematics 2043, DOI 10.1007/978-3-642-24415-5_2,
© Springer-Verlag Berlin Heidelberg 2012

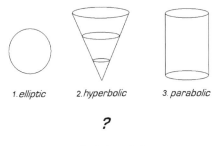

1. elliptic   2. hyperbolic   3. parabolic

?

4. elliptic-hyperbolic

Fig. 2.1 The standard geometry for boundary value problems associated with the four main types of partial differential equation in two space dimensions. In the hyperbolic and parabolic cases, the *vertical axis* represents time. The parabolic case is thus the cartesian product of the disc in case 1 with a *line* representing time. In the hyperbolic case, the disc of case 1 expands over time from an initial point in accordance with Huygens' Principle, to become a cone. (A reflected cone is obtained by expanding "backwards in time.") In the context of elliptic equations in the plane, case 1 is essentially the Riemann Mapping Theorem: see, e.g., the discussion on p. 264 of [15]

$$Lu = \alpha u_{xx} + 2\beta u_{xy} + \gamma u_{yy}, \tag{2.1}$$

where α, β, and γ are given functions of the coordinates $(x, y) \in \Omega$; u is an unknown function of x and y. We can determine the type of the operator L in (2.1) – whatever the lower-order terms – by evaluating the discriminant

$$\Delta(x, y) = \alpha\gamma - \beta^2, \tag{2.2}$$

which will be respectively positive, negative, or zero according to whether (2.1) is elliptic, hyperbolic, or parabolic.

We briefly mention two related consequences of the sign of the discriminant which will be used in the sequel:

Writing the characteristic equation for (2.1) in the form

$$\alpha dy^2 - 2\beta dx dy + \gamma dx^2 = 0, \tag{2.3}$$

we find that the condition $\Delta < 0$ implies that the characteristic lines are real-valued, and solutions to the equation $Lu = 0$ propagate as waves.

Writing the first order system corresponding to the equation $Lu = 0$ in the form

$$A^1 \mathbf{u}_x + A^2 \mathbf{U}_y = 0 \tag{2.4}$$

for $\mathbf{u} = (u_x, u_y)$,

$$A^1 = \begin{pmatrix} \alpha & 0 \\ 0 & 1 \end{pmatrix}$$

and

$$A^2 = \begin{pmatrix} 2\beta & \gamma \\ -1 & 0 \end{pmatrix},$$

we find that the condition $\Delta < 0$ implies that the quadratic form $\tilde{Q} = |A^1 - \lambda A^2|$ associated to the system (2.4) has real roots; c.f. (2.49) and (5.46), below, and [12], Sect. III.2.

If the discriminant is positive on part of Ω and negative elsewhere on Ω, then the equation associated with L is said to be of *mixed elliptic–hyperbolic type*. Recall from Chap. 1 that the curve on which the equation changes type is called the *parabolic curve* and that, borrowing the terminology of fluid dynamics, we will also refer to it as the *sonic curve*. An example is the *Lavrent'ev–Bitsadze equation* [29],

$$Lu = \text{sgn}[y]u_{xx} + u_{yy} = 0, \tag{2.5}$$

for which the parabolic curve is the x-axis.

An alternative approach is to replace the class of differential operators L on a single domain of $\mathbf{R}^2$ with a single differential operator $\mathscr{L}$ on a class of domains. In particular, we may consider the *Laplace–Beltrami operator*

$$\mathscr{L}_g u = \frac{1}{\sqrt{|g|}} \frac{\partial}{\partial x^i} \left(g^{ij} \sqrt{|g|} \frac{\partial u}{\partial x^j} \right), \tag{2.6}$$

on a class of manifolds for which the matrix g_{ij} represents the local metric tensor having determinant g.

In this approach the Lavrent'ev–Bitsadze equation is associated to the Laplace–Beltrami operator on a metric which is Euclidean above the x-axis and Minkowskian below the x-axis. In general, elliptic operators are associated to Riemannian metrics and hyperbolic operators are associated to Lorentzian metrics. In this approach the type of a linear second-order equation is not a function of the associated linear operator at all, as that operator is always the Laplace–Beltrami operator. Instead, the type of the equation is a feature of the metric tensor on the underlying pseudo-Riemannian manifold. Any change in signature which results in a change in sign of the determinant g will change the Laplace–Beltrami operator on the metric from elliptic to hyperbolic type. The Laplace–Beltrami operator on surface metrics for which such a change occurs along a smooth curve will correspond to planar elliptic–hyperbolic operators in local coordinates. Any curve on which the change of type occurs will necessarily represent a singularity of the metric tensor, as g will vanish along that curve. However, none of the terms in the equation itself need blow up on this curve; c.f. (6.12).

According to this point of view, in order to decide which boundary value problems are natural for a second-order linear elliptic–hyperbolic equation on $\mathbf{R}^2$, one should study the geometry of the underlying metric. For example, characteristic lines for the Hodge equations on extended $\mathbf{P}^2$ can be interpreted as polar lines for a chord of the projective disc. This suggests a natural class of boundary value problems, which will be studied in Chap. 6; see also [38]. But this approach has the disadvantage that it fails precisely at the points on which an elliptic–hyperbolic equation is most interesting: the sonic curve. Also, the association of a partial differential equation with a metric having interesting pseudo-Riemannian geometry tends to be the exception rather than the rule in most applications. But see Sect. 6.4.5 for a review of some of these interesting exceptions, and the remarks at the end of Sect. 3 in [39]. Moreover, in the case of (1.4), which has diverse applications, interpreting the equation as the Laplace–Beltrami operator on a metric uncovers symmetries which can be used to apply Noether's Theorem [27]; see Problem 11 of Appendix B. The conservation law which follows from that result can be used to derive a critical uniqueness theorem (Theorem 3.1).

A more robust approach to formulating boundary value problems for elliptic–hyperbolic equations is to declare that the domain should be star-shaped with respect to an appropriate vector field. This approach was developed by Lupo and Payne [33] for equations of Tricomi type. In these notes their ideas will be applied to certain equations of Keldysh type.

2.2   What Is the Right Number of Boundary Conditions?

Probably the simplest requirement for a boundary value problem is that it should have the right number of boundary conditions. But how many boundary conditions that comes out to be is not always obvious for a given differential equation.

Consider, as in [14], [22], the problem of finding an L^2 solution to a simple *ordinary* differential equation

$$x\frac{du}{dx} + u = 0, \tag{2.7}$$

$u = u(x)$, on the interval $-1 \le x \le 1$. (We will discover in Chap. 3 that this is a one-dimensional reduction of a formally self-adjoint Cinquini-Cibrario equation.) Writing

$$\frac{d}{dx}(xu) = 0,$$

we find that $xu = c$, or $u(x) = c/x$, where c is the constant of integration. The solution is in $L^2[-1, 1]$ only if c^2/x^2 is integrable, and that will be the case only if $c = 0$; so this constant of integration can be evaluated without placing any conditions at all at either of the boundary points $x = \pm 1$.

Now consider the problem of finding an L^2 solution to the differential equation

$$-x\frac{du}{dx} + 1 = 0 \tag{2.8}$$

on $[-1, 1]$. Because of the singularity at $x = 0$, the solution can be represented as two improper integrals, one with a lower limit of integration tending to $x = 0$, and the other with an upper limit of integration tending to $x = 0$. Because each integral has an arbitrary constant of integration in its indefinite form, conditions can be prescribed at both $x = -1$ and $x = 1$. The improper integral

$$I^+ \equiv \int_\varepsilon^1 \log^2(x)\,dx$$

is finite as ε tends to zero, and similarly for

$$I^- \equiv \int_{-1}^{-\varepsilon} \log^2 |x|\,dx.$$

We conclude that the solution,

$$u(x) = \log |x| + c,$$

is L^2 on $[-1, 1]$ for the prescribed values of c at the two endpoints.

Friedrichs noticed that this ambiguity in the number of boundary conditions for two linear differential equations of identical order is due to the presence or absence of a singularity. If there were no singularities possible, such first-order ordinary differential equations would be well-posed with a single boundary condition. The boundary condition on (2.7) becomes unnecessary when we require the solution to have a certain degree of smoothness — by imposing an L^2 condition on it. The extra boundary condition in (2.8) arises because we allow a singularity in the domain of the equation.

Applied to mixed elliptic–hyperbolic equations, this kind of thinking will suggest that requiring the solution to extend smoothly across the sonic curve will allow us to remove conditions on some of the boundary arcs, and that if we place conditions on the entire boundary we may still obtain a well-posed problem if we permit the solution to be sufficiently singular.

Physical reasoning may lead to the loss of a boundary condition, for example, in models of flows in nozzles. Alternatively, physical reasoning may suggest more boundary conditions than are warranted mathematically; we will see an example of this in Chap. 4. In that case it is reasonable, based on our motivating examples, to look for solutions to become smoother, in general, as we remove boundary conditions.

We note that methods of complex harmonic analysis can reveal even more choices in the assignment of boundary conditions. For example, L-K. Hua's methods for analyzing linear partial differential equations on Cartan domains imply that the solution of the equation

$$\left(1 - x^2\right) u_{yy} + \left(1 - y^2\right) u_{xx} = 0$$

on the unit square is determined uniquely provided the values for u are prescribed at the four points $x = \pm 1$, $y = \pm 1$ [20].

2.3   What Do We Mean by "Well-posed"?

In order to be *well-posed in the sense of Hadamard*, solutions to a boundary value problem must exist, be unique, and depend continuously on the boundary data. The last condition, which implies that small changes in the boundary data will produce small changes in the solution, is necessary in order to predict the results of actual experiments, in which the boundary data can be measured only up to some finite error. The second condition is also necessary in order to model experiments: if solutions are not unique, then the same experiment, repeated under identical conditions, might yield different results. That would violate the determinism of classical physics.

Here we are concerned mainly with proving the *existence* of solutions, the first step in showing that a boundary value problem is well-posed. This is the mildest requirement for a well-posed boundary value problem, but also the most important. If a solution does not exist, it is usually still possible to run the standard approximation algorithms to generate a numerical result, and nothing in the output file may indicate that any error has been made; but the result will be meaningless nonetheless. Also, the proof of the existence of solutions is usually the most difficult step in showing that a boundary value problem is well-posed. In most cases we will be unable to show the existence of *classical* (*i.e.*, twice continuously differentiable) solutions. Rather we will only be able to show the existence of solutions which lie in a much larger function space.

Such solutions are called *weak*, *distribution*, or *generalized* by different authors representing different traditions or schools of thought. In particular, the term *distribution solution* is often taken to mean a solution which does not lie in a true function space; but our distribution solutions will lie in L^2. There is more possibility for confusion when second-order scalar equations are replaced by first-order systems, as we will do often, as a solution to a first-order system which is only in L^2 corresponds to a solution to the second-order system which is in $H^{1,2}$ – that is, for which the first derivatives are in L^2. (All the differential operators in this text will be either first- or second-order.)

Rather than formulate a general definition of such solutions at the outset, we will define precisely what we mean by a weak, strong, or distribution solution when those objects are introduced. The definitions are summarized in Appendix A.1. We will not use the term "generalized solution" at all, except for a brief remark in Appendix A.1.

The *Dirichlet problem* for a differential equation requires one to find a solution which assumes prescribed values on the boundary of the domain. The *Neumann problem* requires the normal derivative of the solution to assume prescribed values on the boundary. But here again there is some ambiguity, as the Dirichlet problem may only require values to be prescribed on the elliptic part of the domain, on the elliptic part and the non-characteristic hyperbolic part, on the elliptic part and on one of the characteristics, etc. (By the *elliptic part of the boundary*, we mean boundary points on which the differential operator has positive discriminant, and by the *hyperbolic part of the boundary*, points on which it has negative discriminant.) In particular, the *closed* Dirichlet problem refers to the conventional case, in which the values of the solution are prescribed on the entire boundary. In the *open* Dirichlet problem, values of the solution are only prescribed on a proper subset of the boundary. If solutions are prescribed on the elliptic boundary and on the non-characteristic hyperbolic boundary, that open Dirichlet problem is called a *Guderley-Morawetz* problem. If solutions are prescribed on the elliptic part of the boundary and on a characteristic arc, then that open Dirichlet problem is called a *Tricomi* problem. Any of these varieties of Dirichlet problem can be weakly, strongly, or classically well-posed, or well-posed in the sense of distributions, depending on the smoothness of the desired solution. Again, these various problems will be described precisely when they appear in the text.

2.4   Technical Results from Classical Analysis

Here we gather a few technical results that will figure prominently in subsequent chapters. As we remarked in the Preface, no attempt is made in this section to provide a complete preparation in classical or functional analysis for the course.

2.4.1   A Weak Maximum Principle

The following result is true in considerably greater generality than the form in which we prove it; see the Remark following Theorem 3.1 of [17]. It is a *weak* maximum (minimum) principle, which proves that the supremum (infimum) of the function occurs on the boundary, but may also occur in the interior. The *strong* maximum (minimum) principle states that if a maximum (minimum) occurs in the interior, the function is a constant. In the context in which we will apply this result in Chap. 3, either form of the maximum principle would give the same result.

Proposition 2.1 (weak form of a result by E. Hopf; see [44]). *Let $u(x, y)$ be twice-continuously differentiable and satisfy the equation*

$$Lu = a(x, y) u_{xx} + b(x, y) u_x + u_{yy} = 0 \qquad (2.9)$$

on a bounded domain Ω, where $a \geq 0$ and b are bounded. Then u attains both its supremum and infimum on the boundary $\partial \Omega$.

Proof. Suppose that Lw were positive and w attained a maximum at an interior point $(x_0, y_0) \in \Omega$. At that point we would have

$$\det \begin{pmatrix} w_{xx} & w_{xy} \\ w_{xy} & w_{yy} \end{pmatrix} \le 0$$

and

$$w_x = w_y = 0.$$

Because $a \ge 0$, we would also have

$$\det \left[\begin{pmatrix} a & 0 \\ 0 & 1 \end{pmatrix} \begin{pmatrix} w_{xx} & w_{xy} \\ w_{xy} & w_{yy} \end{pmatrix} \right] \le 0,$$

as for any two square matrices A and B,

$$\det AB = (\det A)(\det B).$$

It is also well known (see, e.g., Lemma 8.5 of [49]) that if A and B are *symmetric* square matrices with $A \ge 0$ and $B \le 0$, then $tr(AB) \le 0$. In our case, this means that

$$tr \begin{pmatrix} aw_{xx} & aw_{xy} \\ w_{xy} & w_{yy} \end{pmatrix} = aw_{xx} + w_{yy} \le 0.$$

Because $w_x = 0$ at (x_0, y_0), this contradicts our assumption that

$$Lw = aw_{xx} + bw_x + w_{yy} > 0.$$

We conclude that whenever the operator L of (2.9) is strictly positive, then it satisfies a strong maximum principle, and its argument cannot attain a maximum in the interior of its domain.

Let

$$w = u + \varepsilon e^{\gamma y},$$

for ε and γ positive. Then

$$Lw = Lu + \varepsilon L(e^{\gamma y}) = 0 + \varepsilon \gamma^2 e^{\gamma y} > 0 \ \forall \varepsilon > 0,$$

so any maximum of w must occur on $\partial \Omega$. Letting ε tend to zero, we conclude that

$$\sup_{\Omega} u = \sup_{\partial \Omega} u.$$

Now at a minimum, $w_{xx}w_{yy} - w_{xy}^2 \ge 0$. So if $Lw < 0$, w cannot attain a minimum at an interior point. We obtain

$$\inf_{\Omega} u = \inf_{\partial \Omega} u$$

by defining

$$w = u - \varepsilon e^{\gamma y}$$

and letting ε tend to zero. This completes the proof. □

2.4.2 A Weighted Poincaré Inequality

Let $\mathcal{K}(x, y)$ be a given C^1 function which, unless stated otherwise, will be assumed to be C^1 on Ω. The spaces $L^2(\Omega; |\mathcal{K}|)$ and – if $\mathcal{K}$ does not vanish on a set of positive measure – $L^2(\Omega; |\mathcal{K}^{-1}|)$ consist, respectively, of functions u for which the norm

$$||u||_{L^2(\Omega; |\mathcal{K}|)} = \left(\int\int_\Omega |\mathcal{K}| u^2 dx dy \right)^{1/2}$$

is finite, and functions $u \in L^2(\Omega)$ for which the norm

$$||u||_{L^2(\Omega; |\mathcal{K}|^{-1})} = \left(\int\int_\Omega |\mathcal{K}|^{-1} u^2 dx dy \right)^{1/2}$$

is finite. Analogously, we define the space $H^1(\Omega; \mathcal{K})$ to be the completion of $C^\infty(\Omega)$ with respect to the norm

$$||u||_{H^1(\Omega; \mathcal{K})} = \left[\int\int_\Omega \left(|\mathcal{K}| u_x^2 + u_y^2 + u^2 \right) dx dy \right]^{1/2}, \qquad (2.10)$$

and introduce the space $H_0^1(\Omega; \mathcal{K})$ as the closure of $C_0^\infty(\Omega)$ in this space. The $H_0^1(\Omega; \mathcal{K})$-norm has the form

$$||u||_{H_0^1(\Omega; \mathcal{K})} = \left[\int\int_\Omega \left(|\mathcal{K}| u_x^2 + u_y^2 \right) dx dy \right]^{1/2}, \qquad (2.11)$$

which can be derived from (2.10) via a weighted Poincaré inequality:

Proposition 2.2 (Poincaré–Čanić–Keyfitz). *If $u \in H_0^1(\Omega; \mathcal{K})$, then*

$$||u||_{L^2(\Omega)}^2 \le C(\Omega) ||u||_{H_0^1(\Omega; \mathcal{K})}^2.$$

Proof. (This is essentially Proposition 2 of [8].) It is sufficient to take u to be a continuously differentiable function vanishing on $\partial\Omega$, and to take Ω to be the rectangle

$$R = \{(x, y) | \gamma \le x \le \delta, \beta \le y \le \alpha\}.$$

The boundary condition on u allows us to write

$$u(x, y) = \int_{\beta}^{y} u_t(x, t)\,dt.$$

By the Schwarz inequality,

$$u(x, y) \le \left(\int_{\beta}^{y} dt \right)^{1/2} \left(\int_{\beta}^{y} u_t^2(x, t)\,dt \right)^{1/2} \le C \left(\int_{\beta}^{y} u_t^2(x, t)\,dt \right)^{1/2},$$

where C is a positive constant which depends on R. Squaring both sides (and updating the value of C without changing the notation), we obtain

$$u^2(x, y) \le C \int_{\beta}^{y} u_t^2(x, t)\,dt \le C \int_{\beta}^{\alpha} u_t^2(x, t)\,dt = C \int_{\beta}^{\alpha} u_y^2(x, y)\,dy.$$

If we integrate both sides with respect to y between β and α, we multiply the constant C on the right by a new constant which also depends on R (we will also denote the product of these constants by C), and obtain on the left the integral of u with respect to y between β and α. If we then integrate both sides with respect to x between γ and δ, we obtain

$$\int\int_R u^2(x, y)\,dxdy \le C \int\int_R u_y^2(x, y)\,dxdy \le C \int\int_R \left(|\mathscr{K}| u_x^2 + u_y^2 \right) dxdy.$$

This completes the proof of Proposition 2.2.   □

2.4.3 An Integration-by-Parts Formula

The idea of exploiting a formula like the following in order to prove the uniqueness of solutions to (open) elliptic–hyperbolic boundary value problems is apparently due to Friedrichs, but was first applied by Protter [42, 43]; see also [36]. It is known as the *abc method*. A discussion of this method in the context of equations of Tricomi type can be found in Sect. 12.1 of [15].

Proposition 2.3 (c.f. [41], Sect. II; [37], Proposition 12). *Let*

$$Mu = au + bu_x + cu_y,$$

where $a = const.$, $b = b(x, y)$, *and* $c = c(y)$ *for sufficiently smooth functions* b *and* c *(see the remarks in Sect. 1.4). Let*

$$L_{(\mathscr{K};k)}u = \mathscr{K}(x)u_{xx} + k\mathscr{K}'(x)u_x + u_{yy}, \qquad (2.12)$$

where k is a constant and $\mathcal{K} \in C^2(\Omega)$. If $u \equiv 0$ on $\partial\Omega$, then the L^2-inner product of Mu and $L_{(\mathcal{K};k)}u$ satisfies

$$\left(Mu, L_{(\mathcal{K};k)}u\right) = \oint_{\partial\Omega} \frac{1}{2}\left(\mathcal{K}(x)u_x^2 - u_y^2\right)(c\,dx + b\,dy) + u_x u_y \left(K(x)c\,dy - b\,dx\right)$$

$$+ \int\int_{\Omega} \omega u^2 + \alpha u_x^2 + 2\beta u_x u_y + \gamma u_y^2 \, dx dy,$$

where

$$\omega = (1-k)\frac{a}{2}\mathcal{K}''(x);$$

$$\alpha = \left[\frac{c_y}{2} - \left(a + \frac{b_x}{2}\right)\right]\mathcal{K}(x) + b\left(k - \frac{1}{2}\right)\mathcal{K}'(x);$$

$$2\beta = c\,(k-1)\,\mathcal{K}'(x) - b_y;$$

$$\gamma = \frac{1}{2}\left(b_x - c_y\right) - a.$$

Proof.

$$Mu \cdot Lu = \left(au + bu_x + cu_y\right)\left(\mathcal{K}(x)u_{xx} + u_{yy} + k\mathcal{K}'(x)u_x\right)$$

$$= au\mathcal{K}u_{xx} + auu_{yy} + auk\mathcal{K}'(x)u_x + bu_x\mathcal{K}u_{xx} + bu_xu_{yy}$$

$$+ bu_x^2 k\mathcal{K}'(x) + cu_y\mathcal{K}u_{xx} + cu_yu_{yy} + cu_yk\mathcal{K}'(x)u_x$$

$$\equiv \sum_{i=1}^{9}\tau_i.$$

Taking into account the properties of a, b, c, we have:

$$\tau_1 = au\mathcal{K}u_{xx} = (au\mathcal{K}u_x)_x - au_x^2\mathcal{K} - au\mathcal{K}'(x)u_x$$

$$= (au\mathcal{K}u_x)_x - au_x^2\mathcal{K} - \left(\frac{a}{2}u^2\mathcal{K}'(x)\right)_x + \frac{a}{2}\mathcal{K}''(x)u^2,$$

using the relation $uu_x = (1/2)\left(u^2\right)_x$;

$$\tau_2 = auu_{yy} = \left(auu_y\right)_y - au_y^2;$$

$$\tau_3 = auk\mathcal{K}'(x)u_x = \left(\frac{ak}{2}\mathcal{K}'(x)u^2\right)_x - \frac{ak}{2}\mathcal{K}''(x)u^2,$$

again writing uu_x in terms of the derivative of u^2;

$$\tau_4 = bu_x \mathcal{K} u_{xx} = b\mathcal{K} \frac{1}{2}\left(u_x^2\right)_x = \left(\frac{b}{2}\mathcal{K} u_x^2\right)_x - \frac{b_x}{2}\mathcal{K} u_x^2 - \frac{b}{2}\mathcal{K}'(x)u_x^2;$$

$$\tau_5 = bu_x u_{yy} = \left(bu_x u_y\right)_y - bu_{xy}u_y - b_y u_x u_y = \left(bu_x u_y\right)_y$$
$$-\frac{b}{2}\left(u_y^2\right)_x - b_y u_x u_y = \left(bu_x u_y\right)_y - \left(\frac{b}{2}u_y^2\right)_x + \frac{b_x}{2}u_y^2 - b_y u_x u_y;$$

$$\tau_6 = k\mathcal{K}'(x)bu_x^2;$$

$$\tau_7 = cu_y \mathcal{K} u_{xx} = \left(cu_y \mathcal{K} u_x\right)_x - cu_y \mathcal{K}'(x)u_x - cu_{yx}\mathcal{K} u_x$$
$$= \left(cu_y \mathcal{K} u_x\right)_x - c\mathcal{K}'(x)u_y u_x - \left(\frac{c}{2}\mathcal{K} u_x^2\right)_y + \frac{c_y}{2}\mathcal{K} u_x^2.$$

$$\tau_8 = cu_y u_{yy} = \frac{1}{2}c\left(u_y^2\right)_y = \left(\frac{c}{2}u_y^2\right)_y - \frac{c_y}{2}u_y^2.$$

$$\tau_9 = ck\mathcal{K}'(x)u_x u_y.$$

Integrating over Ω and collecting terms completes the proof. $\square$

The important cases of (2.12) are $k = 1/2$, in which case the operator $L_{(\mathcal{K};k)}$ satisfies a conservation law (Sect. 3.3) and $k = 1$, in which case $L_{(\mathcal{K};k)}$ is self-adjoint (Sect. 3.4). See also the use of this method in Sect. 6.3.3.

2.4.4 Two Theorems from Functional Analysis

We make use in several contexts of the following two theorems, which we state without proof; see, e.g., Sects. 11.1 and 12.4 of [1].

Recall that a mapping $F : X \to \mathbf{R}$ on a real linear space $\mathscr{B}$ is *subadditive* if

$$F(X + Y) \le F(X) + F(Y) \; \forall\, X, Y \in \mathscr{B}.$$

The mapping F is *positive homogeneous* if $\forall a \ge 0$ and all $X \in \mathscr{B}$,

$$F(aX) = aF(X).$$

A subadditive and positive homogeneous mapping is said to be a *sublinear functional*.

Theorem 2.1 (Hahn–Banach Theorem). *Let $\mathscr{M}$ be a subspace of the real linear space $\mathscr{B}$, F a sublinear functional on $\mathscr{B}$, and ℓ a linear functional on $\mathscr{M}$ such that $\forall\, X \in \mathscr{M}$,*

$$\ell(X) \le F(X).$$

Then there exists a linear functional $\mathscr{L}$ defined on all of $\mathscr{B}$ and extending ℓ so that $\forall X \in \mathscr{B}$,

$$\mathscr{L}(X) \le F(X).$$

Theorem 2.2 (Riesz Representation Theorem). *T is a bounded linear functional on a Hilbert space $\mathscr{H}$ if and only if there exists a unique vector $Y \in \mathscr{H}$ such that*

$$T(X) = (X, Y)$$

$\forall X \in \mathscr{H}$, where $(\,,\,)$ denotes the inner product on $\mathscr{H}$.

The uniqueness claim in Theorem 2.2 is obvious; for if there were also a vector $Z \in \mathscr{H}$ for which

$$T(X) = (X, Z)$$

$\forall X \in \mathscr{H}$, then it would follow that

$$(X, Z - Y) = T(X) - T(X) = 0,$$

implying $Z = Y$ by the non-degeneracy of the inner product.

However, this kind of uniqueness, although obvious, is of limited use to us. We will employ Theorem 2.2 in cases for which X results from applying the formal adjoint L^* of a differential operator L to a test function ξ. The boundary values for the argument u of L are prescribed, and uniqueness of the solution to the boundary value problem means that u depends uniquely on the boundary values. There is no reason why the vector Y in Theorem 2.2 should be unique in that sense, and in fact it is not in general.

2.4.5 Divergent Integrals

Due to the reduced regularity of solutions to equations of Keldysh type, integrals of such solutions may fail to converge, even in cases for which the corresponding integrals for equations of Tricomi type do converge. One example concerns the weighted spaces of Sect. 2.4.2, which arise in equations of both types in connection with the closed Dirichlet problem. In this context weighted H^1 can be replaced by unweighted H^1 for solutions of Tricomi-type equations in cases for which this replacement apparently fails for solutions of Keldysh-type equations (see item 1 of the five differences listed in Sect. A.5). Another example concerns integrals representing fundamental solutions, which have singularities in the Keldysh case that are absent in the Tricomi case. Thus, in addition to reviewing the definitions for weighted Sobolev spaces in Sect. 2.4.2, we need to review the corresponding definition for the finite part of a divergent integral. This material will be used (only) in Sect. 3.7.

We follow [11, 19]. Denote by Ω_ε a collection of subdomains of Ω which tends to Ω as ε tends to zero. Let $f(x) \in L^1(\Omega_\varepsilon) \; \forall \varepsilon > 0$, where $x \in \mathbf{R}^n$. Suppose that there are constants $N \in \mathbf{N}$, $C_1, C_2 \ldots C_N, C_* \in \mathbf{R}$ and $\kappa_1 > \kappa_2 > \cdots > \kappa_N \geq 0$ such that

$$\lim_{\varepsilon \to 0} \left(\int_{\Omega_\varepsilon} f(x)dx - C_1\varepsilon^{-\kappa_1} - \cdots - C_N\varepsilon^{-\kappa_N} \right) = C_* \text{ if } \kappa_N > 0$$

and

$$\lim_{\varepsilon \to 0} \left(\int_{\Omega_\varepsilon} f(x)dx - C_1\varepsilon^{-\kappa_1} - \cdots - C_N \log \varepsilon \right) = C_* \text{ if } \kappa_N = 0.$$

Then C_* is the *finite part* of the divergent integral

$$\int_\Omega f(x)dx.$$

We write

$$C_* = F.P. \int_\Omega f(x)dx.$$

For $T \in L^1(\Omega_\varepsilon)$ and any test function $\varphi \in C_0^\infty(\Omega)$, we have

$$\langle F.P.(T), \varphi \rangle = F.P. \langle T, \varphi \rangle.$$

Here $\langle \; \rangle$ is the duality bracket of distribution theory – c.f. expression (3.33), below – and indeed, this expression defines the finite part of T as a distribution.

An example of the finite part of a divergent integral is provided by the principle part of the Cauchy integral; c.f., e.g., Theorem 2.16 of [52]. In this case,

$$\langle T, \varphi \rangle = \left\langle p.v.\frac{1}{x}, \varphi \right\rangle = \lim_{\varepsilon \to 0} \left(\int_{-1}^{-\varepsilon} \frac{1}{x}\varphi(x)dx + \int_\varepsilon^1 \frac{1}{x}\varphi(x)dx \right),$$

where "p.v." denotes the Cauchy principal value. Then [11]

$$T = F.P. \left(\frac{1}{x} \cdot \chi_{(-1,0)}(x) \right) + F.P. \left(\frac{1}{x} \cdot \chi_{(0,1)}(x) \right),$$

where $\chi_A(x)$ is the characteristic function of the set A.

2.5   Symmetric Positive Operators

Consider a system of the form

$$\mathbf{Lu} = \mathbf{f} \tag{2.13}$$

for an unknown vector

$$\mathbf{u} = (u_1(x, y), u_2(x, y)),$$

and a known vector

$$\mathbf{f} = (f_1(x, y), f_2(x, y)),$$

where $(x, y) \in \Omega \subset \mathbf{R}^2$. The operator L satisfies

$$L\mathbf{u} = \begin{pmatrix} \mathscr{K}(x, y) & 0 \\ 0 & -1 \end{pmatrix} \begin{pmatrix} u_1 \\ u_2 \end{pmatrix}_x + \begin{pmatrix} 0 & 1 \\ 1 & 0 \end{pmatrix} \begin{pmatrix} u_1 \\ u_2 \end{pmatrix}_y + \text{ zeroth-order terms.} \quad (2.14)$$

Assume that $\mathscr{K}(x, y)$ is continuously differentiable, negative on a subdomain $\Omega^- \subset\subset \Omega$, positive on $\Omega^+ \subset\subset \Omega$, and zero on a smooth curve separating Ω^+ and Ω^-. If $(f_1, f_2) = (f, 0)$, the components of the vector $\mathbf{u}$ are continuously differentiable, and $u_1 = u_x$, $u_2 = u_y$ for some twice-differentiable function $u(x, y)$, then the first-order system (2.13), (2.14) reduces to a second-order scalar equation having the form

$$\mathscr{K}(x, y) u_{xx} + u_{yy} + \text{ first-order terms} = f.$$

Because the emphasis in this section is on the form of the boundary conditions, the presence or absence of zeroth-order terms in (2.14) will not affect the arguments provided the resulting system is symmetric positive in the sense of (2.16).

A vector $\mathbf{u} \in L^2$ is a *strong solution* of an operator equation of the form (2.13) for $\mathbf{f} \in L^2$, with given boundary conditions, if there exists a sequence $\mathbf{u}^\nu$ of continuously differentiable vectors, satisfying the boundary conditions, for which $\mathbf{u}^\nu$ converges to $\mathbf{u}$ in L^2 and $L\mathbf{u}^\nu$ converges to $\mathbf{f}$ in L^2.

Sufficient conditions for a vector to be a strong solution were formulated by Friedrichs [14]. An operator L associated to an equation of the form

$$L\mathbf{u} = A^1 \mathbf{u}_x + A^2 \mathbf{u}_y + B\mathbf{u}, \quad (2.15)$$

where A^1, A^2, and B are matrices, is said to be *symmetric positive* if the matrices A^1 and A^2 are symmetric and the matrix

$$Q \equiv B^* - \frac{1}{2}\left(A_x^1 + A_y^2\right) \quad (2.16)$$

is a positive-definite matrix operator, where B^* is the symmetrization of the matrix B :

$$B^* = \frac{1}{2}\left(B + B^T\right).$$

The differential equation associated to a symmetric positive operator is also said to be symmetric positive.

Boundary conditions for a symmetric positive equation can be given in terms of a matrix

$$\beta = n_1 A^1_{|\partial\Omega} + n_2 A^2_{|\partial\Omega}, \tag{2.17}$$

where (n_1, n_2) are the components of the outward-pointing normal vector on $\partial\Omega$. The boundary is assumed to be twice-continuously differentiable. Denote by $\mathscr{V}$ the vector space identified with the range of $\mathbf{u}$ in the sense that, considered as a mapping, we have $\mathbf{u} : \Omega \cup \partial\Omega \to \mathscr{V}$. Let $\mathscr{N}(\tilde{x}, \tilde{y})$, $(\tilde{x}, \tilde{y}) \in \partial\Omega$, be a linear subspace of $\mathscr{V}$ and let $\mathscr{N}(\tilde{x}, \tilde{y})$ depend smoothly on $\tilde{x}$ and $\tilde{y}$. A boundary condition $u \in \mathscr{N}$ is *admissible* if $\mathscr{N}$ is a maximal subspace of $\mathscr{V}$ with respect to non-negativity of the quadratic form $(\mathbf{u}, \beta\mathbf{u})$ on the boundary.

A set of sufficient conditions for admissibility is the existence of a decomposition ([14], Sect. 5)

$$\beta = \beta_+ + \beta_-, \tag{2.18}$$

for which:

(1) The direct sum of the null spaces for β_+ and β_- spans the restriction of $\mathscr{V}$ to the boundary

(2) The ranges $\mathscr{R}_\pm$ of $\beta_\pm$ have only the vector $\mathbf{u} = 0$ in common

(3) The matrix $\mu = \beta_+ - \beta_-$ satisfies

$$\mu^* = \frac{\mu + \mu^T}{2} \geq 0 \tag{2.19}$$

These conditions imply that the boundary condition

$$\beta_-\mathbf{u} = 0 \text{ on } \partial\Omega \tag{2.20}$$

is admissible for (2.13) and the boundary condition

$$\mathbf{v}^T \beta_+^T = 0 \text{ on } \partial\Omega \tag{2.21}$$

is admissible for the adjoint problem

$$L^*\mathbf{v} = \mathbf{g} \text{ in } \Omega.$$

The linearity of the operator L and the admissibility conditions on the matrices $\beta_\pm$ imply that both problems possess unique, strong solutions.

Boundary conditions are *semi-admissible* if they satisfy properties (2.19) and (2.20). If $\mathbf{f}$ is in $L^2(\Omega)$ and (2.13) is a symmetric positive equation having semi-admissible boundary conditions, then (2.13) possesses a *weak solution* in the following sense: a vector $\mathbf{u} \in L^2(\Omega)$ such that

$$\int_\Omega (L^*\mathbf{v}) \cdot \mathbf{u} \, d\Omega = \int_\Omega \mathbf{v} \cdot \mathbf{f} \, d\Omega \tag{2.22}$$

for all vectors $\mathbf{v}$ having continuously differentiable components and satisfying (2.21) ([14], Theorem 4.1).

2.5.1 The Friedrichs Identities

The proof of the unique existence of strong solutions to symmetric positive systems having admissible boundary conditions is rather technical, especially if the variety of possible sufficient conditions is taken into account. In addition to Friedrichs' original proof in [14], the classic papers are [30] and a series of papers by Sarason (e.g., [46,47]); also [40] and the extensions to quasilinear equations given in [51] on the basis of ideas by C-H. Gu (see the appendix to [18]). Here we discuss the most elementary features of the theory. Among several possible approaches, we follow that of [22,23]; see also [32].

Let the unknown 2-vector $\mathbf{u} = (u^1, u^2)$ be defined as in (2.13); let $\tilde{A} = (A^1, A^2)$, where A^i, $i = 1, 2$ denotes, as before, a given 2×2 matrix; let G be a given 2×2 matrix. Define the operator

$$\tilde{\nabla}\mathbf{u} = \left(\begin{pmatrix} u_1 \\ u_2 \end{pmatrix}_x, \begin{pmatrix} u_1 \\ u_2 \end{pmatrix}_y \right).$$

Thus $\tilde{A}$ and $\tilde{\nabla}\mathbf{u}$ are both vectors having matrix-valued components. Now we define the class of operators

$$
\begin{aligned}
L\mathbf{u} &= \tilde{A}\tilde{\nabla} \cdot \mathbf{u} + \tilde{\nabla} \cdot (\tilde{A}\mathbf{u}) + G\mathbf{u} \\
&= 2\left(\tilde{A}^1 \mathbf{u}_x + \tilde{A}^2 \mathbf{u}_y\right) + \left(\left(\tilde{A}^1\right)_x + \left(\tilde{A}^2\right)_y + G\right)\mathbf{u} \\
&= A^1 \mathbf{u}_x + A^2 \mathbf{u}_y + B\mathbf{u}
\end{aligned}
\tag{2.23}
$$

for $A^1 = 2\tilde{A}^1$, $A^2 = 2\tilde{A}^2$, and

$$B = \left(\tilde{A}^1\right)_x + \left(\tilde{A}^2\right)_y + G.$$

(In fact the following analysis extends immediately to m-vectors having $r \times r$-matrix-valued components; see Chap. I of [22] for details.)
Then

$$
B - \frac{1}{2}\left(\left(A^1\right)_x + \left(A^2\right)_y\right) = \left(\tilde{A}^1\right)_x + \left(\tilde{A}^2\right)_y + G - \left(\left(\tilde{A}^1\right)_x + \left(\tilde{A}^2\right)_y\right)
$$
$$
= G,
\tag{2.24}
$$

so L is symmetric positive provided the symmetric part, $(G + G^T)/2$, of G is positive definite. Because

$$\beta = \mathbf{n} \cdot \tilde{A},$$

the boundary condition

$$M\mathbf{u} = 0 \text{ on } \partial\Omega,\tag{2.25}$$

where M is a 2×2 matrix, is semi-admissible provided the matrix

$$\mu = M + \beta \tag{2.26}$$

has positive semi-definite symmetric part. (Take $M = -2\beta_-$ in condition (2.20).)

Defining the adjoint operators

$$L^* \mathbf{u} = -\tilde{A} \cdot \tilde{\nabla} \mathbf{u} - \tilde{\nabla} \cdot \left(\tilde{A} \mathbf{u} \right) + G^* \mathbf{u} \tag{2.27}$$

and

$$M^* \mathbf{u} = \left(\mu^* + \beta \right) \mathbf{u}, \tag{2.28}$$

we have the following identities (unless otherwise indicated, the components of $\mathbf{u}$ and $\mathbf{v}$ are assumed to lie in $H^1(\Omega)$):

Lemma 2.1 (First Friedrichs Identity). *If L is symmetric positive, then*

$$(\mathbf{v}, L\mathbf{u}) + (\mathbf{v}, M\mathbf{u})_{|\partial\Omega} = \left(L^* \mathbf{v}, \mathbf{u} \right) + \left(M^* \mathbf{v}, \mathbf{u} \right)_{|\partial\Omega},$$

where $(\ ,\ )$ is the L^2-inner product for 2-vectors.

Proof. Writing $x = x_1$, $y = x_2$, we have

$$(\mathbf{v}, L\mathbf{u}) - \left(L^* \mathbf{v}, \mathbf{u} \right) = \int \int_\Omega \mathbf{v} \cdot \left(\tilde{A} \cdot \tilde{\nabla} \mathbf{u} \right) + \mathbf{v} \cdot \left(\tilde{\nabla} \cdot \left(\tilde{A} \mathbf{u} \right) \right) + \mathbf{v} \cdot G \mathbf{u} \, dx_1 dx_2$$

$$+ \int \int_\Omega \left(\tilde{A} \cdot \tilde{\nabla} \mathbf{v} \right) \cdot \mathbf{u} + \left(\tilde{\nabla} \cdot \left(\tilde{A} \mathbf{v} \right) \right) \cdot \mathbf{u} - G^* \mathbf{v} \cdot \mathbf{u} \, dx_1 dx_2$$

$$= \sum_{i=1}^{2} \int \int \left\{ \mathbf{v} \cdot \left(\tilde{A}^i \mathbf{u}_{x_i} \right) + \mathbf{v} \cdot \left(\tilde{A}^i \mathbf{u}_{x^i} \right) + \left(\tilde{A}^i \mathbf{v}_{x^i} \right) \cdot \mathbf{u} + \left(\tilde{A}^i \mathbf{v} \right)_{x_i} \cdot \mathbf{u} \right\}$$

$$\times \, dx_1 dx_2.$$

In the last line of the preceding equation, the terms in G and G^* have cancelled each other out, as have the terms involving $\nabla \tilde{A}$ (after an integration by parts). Applying the symmetry of the matrices $\tilde{A}^i$, we replace the right-hand side by a divergence and apply Green's Theorem; that is,

$$(\mathbf{v}, L\mathbf{u}) - \left(L^* \mathbf{v}, \mathbf{u} \right) = 2 \sum_{i=1}^{2} \int \int_\Omega \left(\mathbf{v} \cdot \tilde{A}^i \mathbf{u} \right)_{x_i} dx_1 dx_2 = 2 \int_{\partial\Omega} \left(\mathbf{v} \cdot \tilde{A}\mathbf{u} \right) \cdot \mathbf{n} \, dS.$$

Now we apply the definitions to write the integrand of the boundary integral in the form

$$2 \left(\mathbf{v} \cdot \tilde{A} \mathbf{u} \right) \cdot \mathbf{n} = 2 \mathbf{v} \cdot \beta \mathbf{u} = \beta \mathbf{v} \cdot \mathbf{u} + \mathbf{v} \cdot \beta \mathbf{u} = \mu^* \mathbf{v} \cdot \mathbf{u} - \mathbf{v} \cdot \mu \mathbf{u} + \beta \mathbf{v} \cdot \mathbf{u} + \mathbf{v} \cdot \beta \mathbf{u}$$

$$= \mu^* \mathbf{v} \cdot \mathbf{u} + \beta \mathbf{v} \cdot \mathbf{u} - (\mathbf{v} \cdot \mu \mathbf{u} - \mathbf{v} \cdot \beta \mathbf{u})$$

$$= \left(\mu^* + \beta \right) \mathbf{v} \cdot \mathbf{u} - (\mathbf{v} \cdot (M + \beta) \mathbf{u} - \mathbf{v} \cdot \beta \mathbf{u})$$

$$= M^* \mathbf{v} \cdot \mathbf{u} - \mathbf{v} \cdot M \mathbf{u}.$$

This completes the proof. □

Lemma 2.2 (Second Friedrichs Identity). *If L is a symmetric positive operator, then*

$$(\mathbf{u}, L\mathbf{u}) + (\mathbf{u}, M\mathbf{u})_{|\partial\Omega} = (\mathbf{u}, G\mathbf{u}) + (\mathbf{u}, \mu\mathbf{u})_{|\partial\Omega} .$$

Proof. The definitions (2.23), (2.26), (2.27), and (2.28), immediately imply the relations

$$L + L^* = G + G^*$$

and

$$M + M^* = \mu + \mu^*.$$

Now let $\mathbf{v} = \mathbf{u}$ in Lemma 2.1 and divide both sides by two. Rearranging terms, we find that

$$(\mathbf{u}, L\mathbf{u}) + (\mathbf{u}, M\mathbf{u})_{|\partial\Omega} = \frac{1}{2} \left[\left(\mathbf{u}, \left(L + L^* \right) \mathbf{u} \right) + \left(\mathbf{u}, \left(M + M^* \right) \mathbf{u} \right)_{|\partial\Omega} \right]$$

$$= \left(\mathbf{u}, \frac{G + G^*}{2} \mathbf{u} \right) + \left(\mathbf{u}, \frac{\mu + \mu^*}{2} \mathbf{u} \right)_{|\partial\Omega} .$$

An explicit computation shows that

$$\mathbf{u} \cdot \frac{G + G^*}{2} \mathbf{u} = \mathbf{u} \cdot G\mathbf{u},$$

and similarly for $\left(\mu + \mu^* \right)/2$. This completes the proof. □

Lemma 2.3 (Friedrichs Inequality). *Let $\mathbf{u}$ satisfy (2.13) for symmetric positive $\mathbf{L}$, with semi-admissible boundary condition (2.25). If λ_G is the smallest eigenvalue on $\bar{\Omega}$ of the matrix $(G + G^*)/2$, then*

$$||\mathbf{u}||_{L^2} \leq \lambda_G^{-1} ||\mathbf{f}||_{L^2}.$$

Proof. We have, as in the preceding proof,

$$(\mathbf{u}, G\mathbf{u}) = \left(\mathbf{u}, \frac{(G + G^*)}{2} \mathbf{u} \right) \geq (\mathbf{u}, \lambda_G \mathbf{u}) = \lambda_G ||\mathbf{u}||_{L^2}^2,$$

where $\lambda_G \geq 0$ by the condition for symmetric positivity (following (2.24)). Similarly,

$$(\mathbf{u}, \mu\mathbf{u})_{|\partial\Omega} = \left(\mathbf{u}, \left(\frac{\mu + \mu^*}{2}\right)\mathbf{u}\right)_{|\partial\Omega} \geq 0,$$

where $\mu + \mu^* \geq 0$ by the condition for semi-admissibility. Thus

$$\|\mathbf{u}\|_{L^2}^2 \leq \lambda_G^{-1}\left[(\mathbf{u}, G\mathbf{u}) + (\mathbf{u}, \mu\mathbf{u})_{|\partial\Omega}\right] = \lambda_G^{-1}\left[(\mathbf{u}, L\mathbf{u}) + (\mathbf{u}, M\mathbf{u})_{|\partial\Omega}\right],$$

where the final identity on the right follows from Lemma 2.2. Observing that the second term in the sum on the extreme right-hand side vanishes by (2.25), we obtain

$$\|\mathbf{u}\|_{L^2}^2 \leq \lambda_G^{-1}(\mathbf{u}, L\mathbf{u}) = \lambda_G^{-1}(\mathbf{u}, \mathbf{f}) \leq \lambda_G^{-1}\|\mathbf{u}\|_{L^2}\|\mathbf{f}\|_{L^2}.$$

Dividing through by the L^2-norm of $\mathbf{u}$ completes the proof of Lemma 2.3.   □

Theorem 2.3 (Friedrichs Uniqueness Theorem). *If the operator L is symmetric positive, then any C^1 solution to (2.13) satisfying a semi-admissible boundary condition of the form (2.25) is unique.*

Proof. Assume on the contrary that $\mathbf{v}$ and $\mathbf{w}$ are distinct solutions satisfying identical boundary conditions. Then $L(\mathbf{v} - \mathbf{w}) = \mathbf{f} - \mathbf{f} = 0$. Applying Lemma 2.3 to the difference $\mathbf{v} - \mathbf{w}$ implies that $\mathbf{v} = \mathbf{w}$ a.e. Observing that the solutions are C^1 completes the proof.   □

Define the function space

$$\mathscr{V} = C^1(\Omega) \cap \{v | M^*v = 0 \text{ on } \partial\Omega\}.$$

If the components of $\mathbf{f}$ lie in the space $L^2(\Omega)$, then we define a weak solution of (2.13) to be a vector $\mathbf{u}$ such that for every $\mathbf{v} \in \mathscr{V}$ the identity

$$(\mathbf{v}, \mathbf{f}) = (L^*\mathbf{v}, \mathbf{u})$$

is satisfied.

These identities imply:

Theorem 2.4 (Friedrichs Weak Existence Theorem). *If L is symmetric positive and the boundary conditions (2.25) are semi-admissible, then for $\mathbf{f} \in L^2(\Omega)$ there exists a weak solution to (2.13) in the sense of (2.22).*

Proof. Denote by $\mathscr{H}$ the subspace of L^2 consisting of vectors $\mathbf{w}$, having square-integrable components, such that

$$\mathbf{w} = L^*\mathbf{v} \tag{2.29}$$

for $\mathbf{v} \in \mathscr{V}$. Applying Theorem 2.3 with L replaced by L^*, M replaced by M^* and $\mathbf{f}$ replaced by $\mathbf{w}$, we conclude that $\mathbf{v}$ is uniquely determined by $\mathbf{w}$. Moreover, for each $\mathbf{f} \in L^2$ there exists a linear functional $L_\mathbf{f}$ on $\mathscr{H}$ defined by the relation

$$L_{\mathbf{f}}(\mathbf{w}) = (\mathbf{v}, \mathbf{f}). \tag{2.30}$$

Applying Lemma 2.3 to $\mathbf{v}$ (with the same replacements as before), we obtain

$$|(\mathbf{v}, \mathbf{f})| \le ||\mathbf{v}||_{L^2}||\mathbf{f}||_{L^2} \le \lambda_G^{-1}||\mathbf{f}||_{L^2}||\mathbf{w}||_{L^2}.$$

This implies that the linear functional $L_{\mathbf{f}}$ is bounded on the subspace $\mathscr{H}$ of L^2. Applying Theorem 2.1 (Hahn–Banach), we conclude that $L_{\mathbf{f}}$ can be extended to all of L^2. Applying Theorem 2.2 (Riesz Representation), we conclude that there is a vector $\mathbf{u} \in L^2$ for which

$$(\mathbf{w}, \mathbf{u}) = L_{\mathbf{f}}(\mathbf{w}).$$

Applying (2.30) on the right and (2.29) on the left completes the proof. $\qquad\square$

Using the properties of mollifiers in the tangential and normal directions, it is possible to show that if $f \in L^2(\Omega)$ and the boundary $\partial\Omega$ is in C^2, then any weak solution of the equation $Lu = f$ having admissible boundary conditions is also strong. That result has been extended to boundaries having various kinds of corners; see, e.g., [30, 31, 46].

2.5.2 Two Simple Examples

Again following Katsanis [22], we apply the theory of symmetric positive operators to (2.7) and (2.8) discussed in Sect. 2.2.

Multiplying (2.7) through by two, we write the differential operator in the self-adjoint form [22]

$$Lu = x\frac{du}{dx} + \frac{d(xu)}{dx} + u.$$

In the notation of (2.23), L is symmetric positive with $\tilde{A} = x$ and $G = 1$. At the boundary $x = -1$, we have

$$\beta = \mathbf{n} \cdot \tilde{A} = -x.$$

At this boundary choose $\beta_- = 0$, so that $\mu = -x$. In that case

$$M = \mu - \beta = 0.$$

Thus no boundary condition on $u(x)$ is necessary at the left endpoint $x = -1$, as $Mu = 0$ there for every finite value of u. At the right endpoint, $\beta = x$ and the same choice of β_- yields $\mu = x$, again obtaining $M = 0$. Thus we obtain $Mu = 0$ without imposing any boundary condition on $u(x)$ at the right endpoint either. We recover the conclusion that we obtained in Sect. 2.2 by direct integration: no boundary conditions are necessary in order to obtain an admissible boundary condition for an L^2 solution.

Similarly, multiplying both sides of (2.8) by two, we obtain the equation $Lu = -2$, where L can be written in the self-adjoint form

$$Lu = -x\frac{du}{dx} - \frac{d\,(xu)}{dx} + u.$$

In this case, (2.23) implies that $\tilde{A} = -x$, $G = 1$, and this operator, too, is symmetric positive. Choosing $\beta_+ = 0$, so that $\mu = -\beta$, then at either $x = 1$ or $x = -1$,

$$M = \mu - \beta = -2\beta.$$

So conditions must be placed on $u(x)$ at both boundary points, $x = 1$ and $x = -1$, in order to obtain $Mu = 0$ on the boundary. This again confirms the conclusions that we drew from directly integrating the equation.

2.6 Inhomogeneous Boundary Conditions

The boundary value problems considered in this text will tend to have the form

$$Lu = f \text{ in } \Omega, \tag{2.31}$$

$$u = 0 \text{ on } \partial\Omega, \tag{2.32}$$

where L is a linear operator and f an arbitrary function of prescribed smoothness. The presence of the possibly nonzero f allows us to exclude trivial solutions. Moreover, the properties of singular solutions to a homogeneous equation can often be studied by considering weak solutions to the corresponding inhomogeneous equation (particularly for a forcing function in H^{-1}).

The existence of a solution to the system (2.31), (2.32) implies the existence of solutions to the corresponding homogeneous equation with inhomogeneous boundary conditions.

That is, if u is a solution to the boundary value problem (2.31), (2.32) for some linear operator L and every function f in a given function space $\mathscr{S}(\Omega)$, then there is always a solution w to the boundary value problem

$$Lw = 0 \text{ in } \Omega, \tag{2.33}$$

$$w = g \text{ on } \partial\Omega, \tag{2.34}$$

for any given g defined on $\overline{\Omega}$ for which $Lg \in \mathscr{S}(\Omega)$.

In order to see why this assertion must be true, simply choose $f = -Lg$ in (2.31). Then by hypothesis, there is a function u for which

$$Lu = -Lg \text{ in } \Omega.$$

Letting

$$w \equiv u + g,$$

we obtain by linearity,

$$Lw = Lu + Lg = 0 \text{ in } \Omega.$$

Because u satisfies (2.32) on $\partial\Omega$, we conclude that $w = g$ on the boundary. Thus (2.33), (2.34) are satisfied on $\overline{\Omega}$.

Moreover, this argument shows that if u is a unique solution to (2.31), (2.32), then w must be a unique solution to (2.33), (2.34).

2.7   Chapter Appendix: A Survey of Applications

(This section is not necessary for the mathematical development of the course.)

Before launching into the derivations of the technical results, we motivate those discussions by listing some of the things that elliptic–hyperbolic equations are good for. In order to keep the survey to a reasonable length, we consider only those applications that derive from equations having the form

$$\nabla \cdot (\rho(Q)\nabla\varphi) = 0; \qquad\qquad (2.35)$$

here $\varphi(x)$, $x \in \mathbf{R}^n$, is a sufficiently smooth scalar-valued function (although in applications to steady flow with circulation or atmospheric plasma, φ may become multi-valued); $n \geq 2$; $Q = |\nabla\varphi|^2$; $\rho : \mathbf{R}^+ \cup \{0\} \to \mathbf{R}^+$ is a $C^{2,\alpha}$ function of its argument (but a possibly singular function of x); $\alpha \in (0, 1)$.

Taking (2.35) out of divergence form, it can be expressed as the sum

$$\rho(Q)\Delta\varphi - \rho'(Q)\partial_i\varphi\partial_j\varphi\partial_i\partial_j\varphi = 0.$$

2.7.1   Variational Interpretation

Although (2.35) can be derived from physical arguments – e.g., conservation of mass, we will recover it by applying a variational principle to the energy functional

$$E = \frac{1}{2} \int_{\Omega} \int_0^Q \rho(s)ds\, d\Omega,$$

where Ω is a sufficiently regular domain of $\mathbf{R}^n$. Note that if $\rho \equiv 1$, then E is the conventional Dirichlet energy. We have, for arbitrary $\psi \in C_0^\infty(\Omega)$,

$$\delta E = \frac{d}{dt} E\left(\varphi + t\psi\right)_{|t=0} = \frac{1}{2} \int_{\Omega} \rho(Q) \frac{d}{dt} Q\left(\varphi + t\psi\right)_{|t=0} d\Omega$$

$$= \frac{1}{2} \int_{\Omega} \rho(Q) \frac{d}{dt} \left[|\nabla\left(\varphi + t\psi\right)|^2 \right]_{|t=0} d\Omega$$

$$= \frac{1}{2} \int_{\Omega} \rho(Q) \frac{d}{dt} \left(|\nabla\varphi|^2 + 2\langle\nabla\varphi, t\nabla\psi\rangle + t^2 |\nabla\psi|^2 \right)_{|t=0} d\Omega$$

$$= \int_{\Omega} \rho(Q) \langle\nabla\varphi, \nabla\psi\rangle d\Omega = \int_{\Omega} \langle\rho(Q)\nabla\varphi, \nabla\psi\rangle d\Omega$$

$$= \int_{\Omega} d\langle\rho(Q)\nabla\varphi, \psi\rangle - \int_{\Omega} \langle\nabla\cdot[\rho(Q)\nabla\varphi], \psi\rangle d\Omega$$

$$= -\int_{\Omega} \langle\nabla\cdot[\rho(Q)\nabla\varphi], \psi\rangle d\Omega,$$

as ψ has compact support in Ω. (In this context we are using angle brackets to denote point-wise inner product.) At a critical point the variations of E vanish, yielding (2.35) whenever φ is sufficiently smooth.

Equation (2.35) is *quasilinear* – that is, linear in its highest derivative – and changes from elliptic to hyperbolic type along a curve which can be derived by a calculus argument: Taking $n = 2$ for simplicity, we write

$$[\rho(Q)\varphi_x]_x + [\rho(Q)\varphi_y]_y = 0, \tag{2.36}$$

where $Q = \varphi_x^2 + \varphi_y^2$. Observe that

$$[\rho(Q)\varphi_x]_x = \rho(Q)\varphi_{xx} + \rho'(Q)Q_x\varphi_x$$

for

$$Q_x = 2\varphi_x\varphi_{xx} + 2\varphi_y\varphi_{yx}.$$

We therefore have

$$[\rho(Q)\varphi_x]_x = \left[\rho(Q) + 2\varphi_x^2\rho'(Q)\right]\varphi_{xx} + 2\rho'(Q)\varphi_y\varphi_x\varphi_{yx}.$$

Arguing similarly for the second term of (2.36), we can write that equation in the non-divergence form

$$\left[\rho(Q) + 2\varphi_x^2\rho'(Q)\right]\varphi_{xx} + 4\rho'(Q)\varphi_y\varphi_x\varphi_{yx} + \left[\rho(Q) + 2\varphi_y^2\rho'(Q)\right]\varphi_{yy}$$

$$= \alpha\varphi_{xx} + 2\beta\varphi_{xy} + \gamma\varphi_{yy} = 0.$$

This equation is uniformly elliptic provided $\beta^2 - \alpha\gamma < 0$. After some algebraic simplification, this reduces to the strict inequality

$$\rho(Q)\left[\rho(Q) + 2Q\rho'(Q)\right] > 0. \tag{2.37}$$

Because we have assumed that ρ is positive, only the expression in the brackets need be positive. In many standard applications, however – notably, transonic gas dynamics, the density $\rho(Q)$ may tend to zero at some value of Q, a phenomenon known as *cavitation*.

If the strict inequality (2.37) is reversed, then (2.35) becomes hyperbolic; the expression (2.37) degenerates to an identity on the parabolic curve. Because inequality (2.37) depends on φ, it is necessary to actually solve (2.35) in order to write down its parabolic curve in an explicit form, as a curve in the xy-plane.

In applications, (2.35) is often linearized by a hodograph transformation (Sect. 5.3, below). In some of the following applications, the linearized equations are of Keldysh type, while in others, they are of Tricomi type.

2.7.2   A Partial List of Applications

We emphasize that the following applications are by no means exhaustive. For example, they do not include applications to cosmology; see Sect. 6.4.5 for a brief review of those. Nor do they include applications from quantum field theory; one of these is discussed in Problem 20 of Appendix B.1. A large number of elliptic–hyperbolic equations that arise in physics and engineering have no apparent relation to (2.35); see, e.g., [5]. Except where indicated, the following examples are taken to lie on a surface of $\mathbf{R}^3$.

(*i*) If we choose

$$\rho(Q) = \left(1 - \frac{\gamma - 1}{2} Q\right)^{1/(\gamma-1)}, \quad \gamma > 1, \tag{2.38}$$

then (2.35) is the continuity equation for *steady, compressible, irrotational ideal flow* with velocity potential φ and adiabatic constant γ. This model is so thoroughly discussed in the expository literature that we will say nothing more of it until Problem 21 of Appendix B.1. See, e.g., Sect. 2 of [3] for the classical theory, [10] for a recent mathematical treatment, and [9] for some of the physical motivation and numerical results.

(*ii*) If

$$\rho(Q) = \frac{1}{\sqrt{|1 - Q|}}, \quad Q < 1,$$

then (2.35) is the equation for an *elliptic–hyperbolic variational theory for extremal surfaces* in Minkowski 3-space $\mathbf{M}^3$, having graph φ; see Sect. 6.1.

(*iii*) If

$$\rho(Q) = \left|1 \pm \frac{\tau^2}{Q}\right|^{1/2}, \quad Q > 0,$$

Fig. 2.2 Nonlinear
conductivity model for ball
lightning. The lines of current
converge in the neighborhood
of the conducting gas.
Compare to models of a
dielectric sphere in a uniform
external field and to Fig. 5 of
[13]

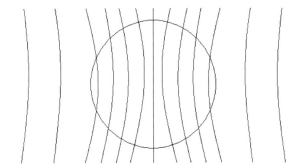

then (2.35) is satisfied by the real $(+)$ or imaginary $(-)$ parts of an eikonal for
a wave propagating in a medium having refractive index τ; see Sect. 5.2.

(*iv*) Choose ρ to represent atmospheric conductivity. Assume that ρ depends
on $Q \equiv |E|^2$, where E is an electric field. Then (2.35) corresponds,
under suitable boundary conditions, to the Finkelstein–Rubinstein *nonlinear
conductivity model for ball lightning* [13]. In this model ball lightning is
represented as a plasma glow discharge surrounded by a region of lower
but rapidly increasing ionization (the so-called *Townsend regime*). Boundary
conditions are

$$E \rightarrow E_0 \text{ as } |x| \rightarrow \infty$$

and

$$\varphi \approx E_0 \cdot x \text{ at } |x| = 1,$$

where E_0 is in the Townsend region surrounding the spherical glow discharge
and $x \in \mathbf{R}^3$.

The problem is classically ill-posed and, indeed, two distinct solutions are
indicated by the physics: a trivial constant solution in which $\varphi = E_0 \cdot x$ and
$\rho(Q) = \rho_0$ in all of space; and a second, nontrivial solution which resembles
the linear problem of a dielectric sphere placed in a uniform electric field
(Fig. 2.2).

If the spherical glow discharge is surrounded by a Townsend regime in
which the field can be represented as a dipole D parallel to an asymptotic
constant field E_0, then this second solution has the explicit form

$$\varphi = E_1 \cdot x, \ \rho(Q) = \rho_1, \ r < 1,$$
$$\varphi = E_0 \cdot x + D \cdot \nabla \left(1/4\pi\varepsilon_0 r\right), \ \rho(Q) = \rho_0, \ r > 1,$$

where r is the radius of the conducting gas discharge and ε_0 is the electric
permittivity of free space.

(*v*) Choosing ρ to have quadratic dependence on the magnetic field (rather
than on the electric field as in the preceding example) leads to models of

ferromagnetism [35]. We do not expect the equations in models (*iv*) and (*v*) to change type.

There are other applications of (2.35) which do not change type, but in which the type degenerates on the parabolic curve. We illustrate these applications with examples (*vi*)–(*ix*), which are extremely well known.

(*vi*) If

$$\rho(Q) = \frac{1}{\sqrt{1+Q}}, \tag{2.39}$$

then (2.35) is the equation for a *nonparametric minimal surface* in $\mathbf{R}^3$ having graph φ; see, e.g., [28]. This equation also arises in gas dynamics, in connection with the Chaplygin approximation; see Sect. 5.6.1.

(*vii*) If

$$\rho(Q) = \frac{1}{\sqrt{1-Q}}, \quad Q < 1, \tag{2.40}$$

then (2.35) is the equation for a *maximal spacelike hypersurface* in Minkowski 3-space $\mathbf{M}^3$, having graph φ; see, e.g., [7] and Sect. 6.1.

(*viii*) If $n = 3$ and $\rho(Q)$ is a suitable power of Q, then we obtain from (2.35) a model of *non-Newtonian pseudo-plastic flow*; see, e.g., [2].

(*ix*) If we replace the scalar φ by a vector $\mathbf{A}$ having components A_i, then we can replace (2.35) by the analogous system

$$\partial_j \left(\rho(Q) \partial_j A_i \right) = 0, \tag{2.41}$$

where $Q = |\nabla \times \mathbf{A}|^2$, and i and j run from 1 to n. If $n = 4$ and $\rho(Q)$ is taken to be the minimal surface density (2.39) or the maximal spacelike hypersurface density (2.40), then one obtains from (2.41), respectively, the Euclidean or Lorentzian *Born–Infeld models for electromagnetism*:

$$\rho(Q) = \frac{1}{\sqrt{1 \pm Q}},$$

where $\mathbf{A}$ is an electromagnetic vector potential. This model was originally introduced in [6] to remove the fundamental singularity of conventional electromagnetic theory, but has attained new interest in connection with brane theories; for recent treatments see [16, 48, 53]. See also [26] for a related model.

(*x*) Whereas the application of (2.35) to gas dynamics is extremely well known, its application to hydrodynamics is less widely known, although a relation between compressible gas dynamics and shallow hydrodynamics was already noticed (for the case of one space dimension) as early as 1932 [45].

Consider steady, inviscid, hydrodynamic flow in a shallow channel. Write the flow velocity in components (u, v, w), where u is the horizontal component in the x-direction, v is the horizontal component in the y-direction, and w is the component in the (vertical) z-direction. Impose initial conditions

Fig. 2.3 Components of
velocity for shallow
hydrodynamic flow governed
by a hydrostatic law; $\mathbf{u} = u\hat{\mathbf{i}}$,
$\mathbf{v} = v\hat{\mathbf{j}}$

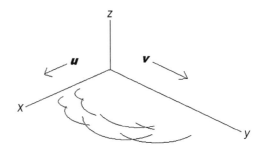

under which w is zero at time $t = 0$. Because the flow is shallow, we suppose
that the component of acceleration of water particles in the z-direction has
negligible effect on pressure. The result of this *hydrostatic law* is that w
remains zero for all subsequent times and the horizontal velocity components
u and v are independent of the z-coordinate (Fig. 2.3). Because the flow is
steady, the velocity components are also independent of t. Moreover, we
assume that the flow is irrotational, so that a flow potential exists locally.

We write the equation of continuity in the form

$$0 = [h\,(x, y)\,u\,(x, y)]_x + [h\,(x, y)\,v\,(x, y)]_y = h_x u + h u_x + h_y v + h v_y, \tag{2.42}$$

where $h(x, y)$ represents the depth of the channel at the point (x, y). This
corresponds to (2.35) with $\rho = h$. Bernoulli's Law can be written in the form

$$h = \frac{K - \left(u^2 + v^2\right)}{2g}, \tag{2.43}$$

where K is a positive constant and g is the acceleration due to gravity ("still
waters run deep"). Substituting (2.43) into (2.42) yields

$$-\frac{(uu_x + vv_x)}{g} u + \frac{K - \left(u^2 + v^2\right)}{2g}\left(u_x + v_y\right) - \frac{\left(uu_y + vv_y\right)}{g} v = 0.$$

Collecting terms,

$$\left(\frac{K - 3u^2 - v^2}{2g}\right) u_x - \frac{\left(u_y + v_x\right) uv}{g} + \left(\frac{K - 3v^2 - u^2}{2g}\right) v_y = 0.$$

But (2.43) implies that

$$K = 2gh + u^2 + v^2,$$

so

$$\left(2gh - 2u^2\right) u_x - 2\left(u_y + v_x\right) uv + \left(2gh - 2v^2\right) v_y = 0.$$

Writing $c^2 = gh$ and dividing by two, this is

$$\left(c^2 - u^2\right) u_x - \left(u_y + v_x\right) uv + \left(c^2 - v^2\right) v_y = 0. \qquad (2.44)$$

In the neighborhood of any point of flow, we can express the velocity components in terms of the flow potential: $u = -\varphi_x$ and $v = -\varphi_y$. We find that (2.44) is locally equivalent to a scalar equation of second order, which is elliptic for squared flow speeds satisfying

$$u^2 + v^2 < c^2 \qquad (2.45)$$

and hyperbolic for squared flow speeds satisfying the reverse inequality.

Flows satisfying inequality (2.45) are said to be *subcritical, tranquil,* or *streaming.* If the reverse inequality is satisfied, the flow is said to be *supercritical* or *shooting.* Otherwise, the flow is *critical.* Shooting flow characterizes turbulent shallow water. Examples include tidal bores which occur, for example, in the English rivers Severn and Trent, the French river Seine near Caudebec-en-Caux, and the Chinese river Tsien–Tang.

Shooting flow is more commonly observed in rapids, in which the depth of a channel dramatically decreases, resulting in a steep increase in flow velocity due to the Law of Mass Conservation (Fig. 2.4).

The *Froude number,*

$$\mathscr{F} \equiv \sqrt{\frac{u^2 + v^2}{gh}},$$

provides a simple characterization of tranquil flow ($\mathscr{F} < 1$) versus shooting flow ($\mathscr{F} > 1$).

For more detailed discussions of these ideas, see Sect. 2.7 of [21] and Sect. 10.12 of [50].

(*xi*) Equations of the form (2.35) can be used to construct models of traffic flow. We will consider only those models in which the equation changes from elliptic to hyperbolic type along a smooth curve [4, 24].

Consider a two-lane highway in which the cars in each of the lanes travel in opposite directions. The traffic in the two lanes can be approximated

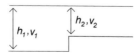

Fig. 2.4 The formation of river rapids: the height of the channel is denoted by h_1 and h_2. The corresponding flow velocities are denoted by v_1 and v_2, respectively. Because $h_1 > h_2$, we have $v_1 < v_2$

by continuous quantities having densities $p(x,t)$ and $q(x,t)$, with corresponding velocities $u(x,t)$ and $v(x,t)$, $x \in \mathbf{R}$, $t \in \mathbf{R}^+$. This *continuum* approximation is analogous to the gas-dynamic and hydrodynamic models, in which molecular motions are averaged into continuous density functions. If we consider a stretch of highway between entrances and exits and do not allow U-turns, then each lane obeys an analogue of the law of mass conservation.

In the applications to gas dynamics and hydrodynamics, (2.35) is the steady-state case of an evolution equation having the form

$$\rho_t + \nabla \cdot (\rho u) = 0,$$

and that representation is also shared with the traffic model. In that model there is one space dimension and ρ depends on x and t. The presence of two lanes results in a system of two equations, having the form

$$p(x,t)_t + [p(x,t)\, u(x,t)]_x = 0,$$
$$q(x,t)_t + [q(x,t)\, v(x,t)]_x = 0. \tag{2.46}$$

The crucial hypothesis in the model is that the two lanes of traffic interact in some way. That is, the speed of the traffic in one direction is affected not only by the density of the traffic in that lane, but also, to some degree, by the density of the traffic in the opposing lane. In that case the two equations of (2.46) are *coupled*. Coupled systems arise in certain models of two-phase incompressible flow; c.f. (8) and (9) of [25]. Coupled systems related to (2.35) also arise – in higher dimensions – in the Born–Infeld model, application (*ix*), above; c.f. (2.38) and (2.39) of [53].

As an example, choose

$$u = 1 - p - \beta q \tag{2.47}$$

and

$$v = -(1 - q - \beta p) \tag{2.48}$$

with $0 < \beta << 1$. The size of the number β should be taken as a (crude) measure of the quantity of interaction between the two lanes. In distinction to the preceding applications, in this model the flow velocities u and v are known functions of the unknown densities p and q.

Recall from Sect. 2.1 that a first-order system of the form (2.4) will be hyperbolic on the region on which its associated quadratic form $\tilde{Q}$ has real roots. Substituting (2.47) and (2.48) into the system (2.46), we write the resulting coupled system in the form

$$A^1 \mathbf{U}_t + A^2 \mathbf{U}_x = 0,$$

where $\mathbf{U} = (p, q)^T$, A^1 is the 2×2 identity matrix, 0 is the 2×2 zero matrix, and

$$A^2 = \begin{pmatrix} 1 - 2p - \beta q & -\beta p \\ \beta q & -(1 - 2q - \beta p) \end{pmatrix}.$$

Solving the eigenvalue equation

$$\left| A^1 - \lambda A^2 \right| = 0, \tag{2.49}$$

we find that the system (2.46)–(2.48) changes from elliptic to hyperbolic type on a region in the pq-plane having area depending on β; c.f. the discussion on p. 194 of [4]. So this nonlinear traffic model is of mixed elliptic–hyperbolic type.

2.7.3   Remark

It is natural to ask whether there is a relation between these diverse choices of mass density $\rho(Q)$. The answer is that in many cases they can be related by Bäcklund transformations. This approach was developed by Magnanini and Talenti [34] in connection with a dual choice of densities in non-geometrical optics. In Chap. 5 we will review their work (Sects. 5.2 and 5.4), as well as a coordinate-independent extension of their method to a large class of mass density functions (Sect. 5.6).

References

1. Bachman, G., Narici, L.: Functional Analysis. Academic Press, San Diego (1966)
2. Beirão de Veiga, H.: On non-Newtonian p-fluids. The pseudo-plastic case. J. Math. Anal. Appl. **344**, 175–185 (2008)
3. Bers, L.: Mathematical Aspects of Subsonic and Transonic Gas Dynamics. Wiley, New York (1958)
4. Bick, J.H., Newell, G.F.: A continuum model for two-dimensional traffic flow. Quart. Appl. Math. **18**, 191–204 (1960)
5. Bîlă, N.: Application of symmetry analysis to a PDE arising in the car windshield design. SIAM J. Appl. Math. **65**, 113–130 (2004)
6. Born, M., Infeld, L.: Foundation of a new field theory. Proc. R. Soc. London Ser. A **144**, 425–451 (1934)
7. Calabi, E.: Examples of Bernstein problems for some nonlinear equations, Proceedings of the Symposium on Global Analysis, pp. 223–230. University of California at Berkeley (1968)
8. Čanić, S., Keyfitz, B.: An elliptic problem arising from the unsteady transonic small disturbance equation J. Diff. Equations **125**, 548–574 (1996)
9. Chapman, C.J.: High Speed Flow. Cambridge University Press, Cambridge (2000)

10. Chen, G-Q., Feldman, M.: Multidimensional transonic shocks and free boundary problems for nonlinear equations of mixed type. J. Amer. Math. Soc. **16**, 461–494 (2003)

11. Chen, S-X.: The fundamental solution of the Keldysh type operator, Science in China, Ser. A: Mathematics **52**, 1829–1843 (2009)

12. Courant, R., Hilbert, D.: Methods of Mathematical Physics, vol. 2. Wiley-Interscience, New York (1962)

13. Finkelstein, D., Rubinstein, J.: Ball lightning. Phys. Rev. **135**, A390–A396 (1964)

14. Friedrichs, K.O.: Symmetric positive linear differential equations. Commun. Pure Appl. Math. **11**, 333–418 (1958)

15. Garabedian, P.: Partial Differential Equations. American Mathematical Society, Providence (1998)

16. Gibbons, G.W.: Born-Infeld particles and Dirichlet p-branes. Nucl. Phys. B **514**, 603–639 (1998)

17. Gilbarg, D., Trudinger, N.S.: Elliptic Partial Differential Equations of Second Order. Springer, Berlin (1983)

18. Gu, C.: On partial differential equations of mixed type in n independent variables. Commun. Pure Appl. Math. **34**, 333–345 (1981)

19. Hadamard, J.: Le problème de Cauchy et les équations aux dérivées partielles linéaires hyperboliques. Hermann, Paris (1932)

20. Hua, L.K.: Geometrical theory of partial differential equations. In: Chern, S.S., Wen–tsün, W. (eds.) Proceedings of the 1980 Beijing Symposium on Differential Geometry and Differential Equations, pp. 627–654. Gordon and Breach, New York (1982)

21. Johnson, R.S.: A Modern Introduction to the Mathematical Theory of Water Waves. Cambridge University Press, Cambridge (1997)

22. Katsanis, T.: Numerical techniques for the solution of symmetric positive linear differential equations. Ph.D. thesis, Case Institute of Technology (1967)

23. Katsanis, T.: Numerical solution of symmetric positive differential equations. Math. Comp. **22**, 763–783 (1968)

24. Keyfitz, B.L.: Hold that light! Modeling of traffic flow by differential equations. In: Hardt, R., Forman, R. (eds.) Six Themes on Variation, pp. 127–153. American Mathematical Society, Providence (2005)

25. Keyfitz, B.L.: Mathematical properties of nonhyperbolic models for incompressible two-phase flow, e-print (2000). http://www.math.osu.edu/b̃keyfitz/blkp.html. Cited 1 Aug 2011

26. Kong, D-X.: A nonlinear geometric equation related to electrodynamics. Europhys. Lett. **66**, 617–623 (2004)

27. Kosmann-Schwarzbach, Y.: The Noether theorems: Invariance and conservation laws in the twentieth century. Sources and studies in the History of Mathematics and Physical Sciences. Springer, Berlin (2010)

28. Kreyszig, E.: On the theory of minimal surfaces. In: Rassias, Th.M. (ed.) The Problem of Plateau: A Tribute to Jesse Douglas and Tibor Radó, pp. 138–164. World Scientific, Singapore (1992)

29. Lavrent'ev, M.A., Bitsadze, A.V.: On the problem of equations of mixed type [in Russian], Doklady Akad. Nauk SSSR (n.s.) **70**, 373–376 (1950)

30. Lax, P.D., Phillips, R.S.: Local boundary conditions for dissipative symmetric linear differential operators. Commun. Pure Appl. Math. **13**, 427–455 (1960)

31. Lin, C.S.: The local isometric embedding in $\mathbf{R}^3$ of a 2-dimensional Riemanian manifolds with Gaussian curvature changing sign cleanly. Commun. Pure Appl. Math. **39**, 867–887 (1986)

32. Liu, J-L.: A finite difference method for symmetric positive operators. Math. of Computation **62**, 105–118 (1994)

33. Lupo, D., Payne, K.R.: Critical exponents for semilinear equations of mixed elliptic-hyperbolic and degenerate types. Commun. Pure Appl. Math. **56**, 403–424 (2003)

34. Magnanini, R., Talenti, G.: On complex-valued solutions to a 2D eikonal equation. Part one: qualitative properties. Contemporary Math. **283**, 203–229 (1999)

35. Milani, A., Picard, R.: Decomposition and their application to non-linear electro- and magnetostatic boundary value problems. In: Hildebrandt, S., Leis, R. (eds.) Partial Differential Equations and Calculus of Variations, Lecture Notes in Mathematics, vol. 1357, pp. 370–340. Springer, Berlin (1988)

36. Morawetz, C.S.: A weak solution for a system of equations of elliptic-hyperbolic type. Commun. Pure Appl. Math. **11**, 315–331 (1958)

37. Otway, T.H.: Energy inequalities for a model of wave propagation in cold plasma. Publ. Mat. **52**, 195–234 (2008)

38. Otway, T.H.: Variational equations on mixed Riemannian-Lorentzian metrics. J. Geom. Phys. **58**, 1043–1061 (2008)

39. Payne, K.R.: Solvability theorems for linear equations of Tricomi type. J. Mat. Anal. Appl. **215**, 262–273 (1997)

40. Phillips, R.S., Sarason, L.: Singular symmetric positive first order differential operators. J. Math. Mech. **8**, 235–272 (1966)

41. Pilant, M.: The Neumann problem for an equation of Lavrent'ev-Bitsadze type. J. Math. Anal. Appl. **106**, 321–359 (1985)

42. Protter, M.H.: Uniqueness theorems for the Tricomi problem. I. J. Rat. Mech. Anal. **2**, 107–114 (1953)

43. Protter, M.H.: Uniqueness theorems for the Tricomi problem, II. J. Rat. Mech. Anal. **4**, 721–732 (1955)

44. Protter, M.H., Weinberger, H.F.: Maximum Principles in Differential Equations. Springer, Berlin (1984)

45. Riabouchinsky, D.: Sur l'analogie hydraulique des mouvements d'un fluide compressible. C. R. Academie des Sciences, Paris **195**, 998 (1932)

46. Sarason, L.: On weak and strong solutions of boundary value problems. Commun. Pure Appl. Math. **15**, 237–288 (1962)

47. Sarason, L.: Elliptic regularization for symmetric positive systems. J. Math. Mech. **16**, 807–827 (1967)

48. Sibner, L.M., Sibner, R.J., Yang, Y.: Generalized Bernstein property and gravitational strings in Born–Infeld theory. Nonlinearity **20**, 1193–1213 (2007)

49. Smoller, J.: Shock Waves and Reaction-Diffusion Equations. Springer, Berlin (1983)

50. Stoker, J.J.: Water Waves. Interscience, New York (1987)

51. Tso, K. (Chou, K-S.): Nonlinear symmetric positive systems. Ann. Inst. Henri Poincaré **9**, 339–366 (1992)

52. Wilde, I.F.: Distribution Theory (Generalized Functions). e-Notes. http://homepage.ntlworld. cpm/ivan.wilde/notes/gf/gf.pdf. Cited 2 Aug 2011

53. Yang, Y.: Classical solutions in the Born-Infeld theory. Proc. R. Soc. Lond. Ser. A **456**, 615–640 (2000)

Chapter 3
The Equation of Cinquini-Cibrario

3.1 Very Brief Historical Remarks

Although the study of mixed elliptic–hyperbolic equations goes back at least to Riemann's computation of the Laplacian in toroidal coordinates (c.f. [46] or p. 461, (B) of [7]), the first systematic study of well-posedness for boundary value problems appears to be the memoir by Tricomi [49]. Reasoning from purely mathematical assumptions, Tricomi studied the equation

$$yu_{xx} + u_{yy} = 0, \qquad (3.1)$$

where $u = u(x, y)$. This equation is typically defined on a domain $\Omega \subset\subset \mathbf{R}^2$ bounded by a smooth Jordan curve γ in the upper half plane and the characteristic lines

$$x + \frac{2}{3}(-y)^{3/2} = 1, \ x > 0, \qquad (3.2)$$

$$x - \frac{2}{3}(-y)^{3/2} = -1, \ x < 0, \qquad (3.3)$$

in the lower half-plane (Fig. 3.1). Tricomi proved that a unique solution to (3.1) exists on Ω provided the values of $u(x, y)$ are smoothly prescribed on the elliptic arc γ and the characteristic line (3.3).

Tricomi further claimed that any sufficiently smooth equation having the form

$$\alpha(x, y) u_{xx} + 2\beta(x, y) u_{xy} + \gamma(x, y) u_{yy} + \text{lower order} = 0,$$

for which the discriminant $\beta^2 - \alpha\gamma$ changes sign along a smooth curve, is locally equivalent to (3.1) plus lower order terms. This erroneous statement (which is repeated, for example, on pp. 81 and 82 of the classic text [11]) was corrected in the early 1930s by Cinquini-Cibrario, who studied the equation [18]

T.H. Otway, *The Dirichlet Problem for Elliptic-Hyperbolic Equations of Keldysh Type*, Lecture Notes in Mathematics 2043, DOI 10.1007/978-3-642-24415-5_3, © Springer-Verlag Berlin Heidelberg 2012

Fig. 3.1 A typical Tricomi domain. The elliptic region is bounded by the curve γ and lies above the x-axis; the hyperbolic region is bounded by the characteristic lines Γ_1, Γ_2 and lies below the x-axis. The hyperbolic boundary is the graph of the equation $y = -\{(3/2)\,[(1-x)]\}^{2/3}$

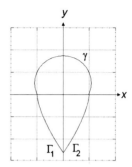

$$x u_{xx} + u_{yy} = 0. \tag{3.4}$$

Her investigations led to the discovery of a small number of canonical forms, slightly generalizing (3.1) and (3.4), to which any linear second-order elliptic–hyperbolic equation is reducible near a point of $\mathbf{R}^2$ (see [17], and Chap. 1 of [12]).

It was shortly after Cinquini-Cibrario abandoned her study of elliptic–hyperbolic equations in favor of other areas of analysis that von Kármán and Frankl' [22] independently drew attention to the physical significance of equations closely related to (3.1), having the forms

$$y u_{xx} - u_{yy} = 0 \tag{3.5}$$

and

$$\mathscr{K}(y) u_{xx} + u_{yy} = 0, \tag{3.6}$$

where

$$\mathscr{K}(0) = 0 \tag{3.7}$$

and

$$y \mathscr{K}(y) > 0 \text{ for } y \neq 0. \tag{3.8}$$

Such equations arise as linear approximations to the quasilinear *continuity equation* for steady ideal potential flow in the plane (Sect. 2.7.2, application (*i*)). The approximation is valid near the sonic curve, at which the character of the continuity equation changes from elliptic to hyperbolic type. Equations having the form (3.6) – possibly including lower-order terms – and satisfying conditions (3.7) and (3.8) are said to be of Tricomi type (c.f. (1.1)). Despite the importance of Cinquini-Cibrario's contribution, equations having the form

$$\mathscr{K}(x) u_{xx} + u_{yy} + \text{lower-order terms} = 0, \tag{3.9}$$

satisfying (3.7) and

$$x \mathscr{K}(x) > 0 \text{ for } x \neq 0, \tag{3.10}$$

are said to be of *Keldysh type* (c.f. (1.2)), in honor of the mathematician who studied the degeneration of ellipticity in such equations in the early 1950s [30]. We will use this established terminology, except that (3.4) will be called *Cinquini-Cibrario's equation,* rather than the *Keldysh equation* as in, e.g., [16].

3.2 Transformation to Canonical Form

In considering canonical forms for elliptic–hyperbolic equations, we ignore lower-order terms as they do not significantly affect the analysis. Lower-order terms are, however, often crucial to the existence of solutions to equations having the form (3.9), as will be shown.

Suppose our equation originally has the form

$$\alpha(x, z) \varphi_{xx} + \gamma(x, z) \varphi_{zz} = 0, \tag{3.11}$$

where γ is nonvanishing on the domain. (All the models considered in the sequel will have highest-order terms of this form, either in cartesian or polar coordinates. The coefficient of φ_{xz} can be set to zero by a standard coordinate change; see Sect. 1.2 of [12].) Perform the coordinate transformation $(x, z) \rightarrow (\xi(x, z), \eta(x, z))$, where

$$\xi = \alpha(x, z).$$

In these coordinates,

$$\alpha\varphi_{xx} + \gamma\varphi_{zz} = \left(\xi\xi_x^2 + \gamma\xi_z^2\right)\varphi_{\xi\xi}$$
$$+2\left(\xi\xi_x\eta_x + \gamma\xi_z\eta_z\right)\varphi_{\xi\eta} + \left(\xi\eta_x^2 + \gamma\eta_z^2\right)\varphi_{\eta\eta}. \tag{3.12}$$

If the transformation $(x, z) \rightarrow (\xi, \eta)$ is to be nonsingular, its Jacobian must be nonvanishing, *i.e.,*

$$\xi_x\eta_z - \xi_z\eta_x \neq 0. \tag{3.13}$$

In addition, we want the coefficients of the cross term $\varphi_{\xi\eta}$ to be zero in the new coordinates; so, taking into account (3.12), we impose the condition that

$$\xi\xi_x\eta_x + \gamma\xi_z\eta_z = 0. \tag{3.14}$$

It is easy for the two first derivatives of η to satisfy (3.13) and (3.14) simultaneously for given ξ and γ.

Either

i) ξ and ξ_z never vanish simultaneously on the sonic curve, or else
ii) There exist one or more points (x, z) on the sonic curve at which

$$\xi(x, z) = \xi_z(x, z) = 0. \tag{3.15}$$

Because $\xi = \alpha(x, z)$ and γ does not vanish, case *ii)* is equivalent to the case in which the characteristics of (3.11) are tangent to the sonic curve; c.f. (2.3).

In case *i)*, in order for ξ to vanish we would need, given condition (3.14) and the assumption that γ is positive, the additional condition $\eta_z = 0$. But if ξ and η_z both vanish, then the coefficient of $\varphi_{\eta\eta}$ in (3.12) also vanishes. That is,

$$\xi \eta_x^2 + \gamma \eta_z^2 = 0 \tag{3.16}$$

whenever $\xi = 0$.

Applying (3.14) to (3.12), we obtain from (3.11) the equation

$$\varphi_{\xi\xi} + \frac{\xi \eta_x^2 + \gamma \eta_z^2}{\xi \xi_x^2 + \gamma \xi_z^2} \varphi_{\eta\eta} = 0. \tag{3.17}$$

The denominator in the coefficient of $\varphi_{\eta\eta}$ cannot be zero: ξ and ξ_z cannot vanish simultaneously, and if ξ_x vanishes, then ξ_z must be nonzero in order to preserve condition (3.13). The numerator in the coefficient of $\varphi_{\eta\eta}$ vanishes whenever $\xi = 0$, by the arguments leading to (3.16). Thus (3.17) is an equation having the form

$$\varphi_{\xi\xi} + \mathcal{K}(\xi, \eta)\, \varphi_{\eta\eta} = 0, \tag{3.18}$$

where $\mathcal{K}(\xi, \eta) = 0$ when $\xi = 0$, an equation of *Tricomi type*.

In case *ii)*, condition (3.13) prevents η_z from vanishing when ξ_z vanishes. Thus in case *ii)* we obtain from (3.11), (3.12), and (3.14) an equation of the form

$$\frac{\xi \xi_x^2 + \gamma \xi_z^2}{\xi \eta_x^2 + \gamma \eta_z^2} \varphi_{\xi\xi} + \varphi_{\eta\eta} = 0, \tag{3.19}$$

where the numerator in the coefficient of $\varphi_{\xi\xi}$, but not the denominator, is zero at one or more points on the sonic curve when ξ is zero there. That is, (3.19) is an equation of the form

$$\mathcal{K}(\xi, \eta)\, \varphi_{\xi\xi} + \varphi_{\eta\eta} = 0, \tag{3.20}$$

where $\mathcal{K}(\xi, \eta) = 0$ when $\xi = 0$, an equation of *Keldysh type*. That is what we would expect from our interpretation of case *ii)* as requiring that the characteristics of (3.11) be tangent to the sonic curve.

This line of reasoning establishes the two canonical forms for equations originally having the form (3.11). See Sect. 1.2 of [12, 17], Sect. 1 of [40], and (75)–(78) of [52] for similar arguments.

3.3 A Closed Dirichlet Problem Which Is Classically Ill-Posed

In 1956, Morawetz [38] proved the uniqueness of sufficiently smooth solutions to *open* Dirichlet problems, having data prescribed on only part of the boundary, for certain mixed elliptic–hyperbolic equations of Tricomi type. That result implied that the closed Dirichlet problem is over-determined for sufficiently smooth solutions of such equations.

Morawetz's result was later extended to a large class of boundary value problems for Tricomi-type equations, by Manwell ([36] and Sect. 16 of [37]) and by Morawetz herself [39]. In this section we extend the result to a large class of equations of Keldysh type.

For given $\mathscr{K}(x)$, define constants a, b, d, and m, where $m < a \leq 0 < d$ and $b > 0$. Consider the domain $\mathscr{D}$ formed by the line segments

$$\mathscr{L}_1 = \{(x, y)\,|a \leq x \leq d, y = -b\};$$

$$\mathscr{L}_2 = \{(x, y)\,|x = d, -b \leq y \leq b\};$$

$$\mathscr{L}_3 = \{(x, y)\,|a \leq x \leq d, y = b\};$$

the characteristic line Γ_1 joining the points $(m, 0)$ and $(a, -b)$; and the characteristic line Γ_2 joining the points $(m, 0)$ and (a, b); see Fig. 3.2.

We consider equations having the form

$$Lu \equiv \mathscr{K}(x)u_{xx} + u_{yy} + \frac{\mathscr{K}'(x)}{2}u_x = 0, \tag{3.21}$$

where $\mathscr{K}$ satisfies conditions (3.7) and (3.10). We assume for convenience that $\mathscr{K}$ is C^1, and monotonic on the hyperbolic region, but the result clearly extends to weaker hypotheses on $\mathscr{K}$. For example, the monotonicity hypothesis on $\mathscr{K}$ is imposed only in order to simplify the graphs of the characteristic lines, which in turn simplifies the proof of Theorem 3.1 in the hyperbolic region. See also the discussion of the type-change function (3.32), below.

As an example, choose $\mathscr{K}(x) = x^{2k_0-1}$ for $k_0 \in \mathbf{Z}^+$. The operator L under this choice of type-change function is an analogue, for equations of Keldysh type,

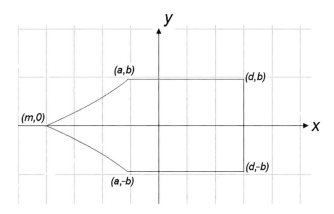

Fig. 3.2 The domain $\mathscr{D}$

of the well known *Gellerstedt operator* [24] for equations of Tricomi type. Other roughly equivalent examples are the polar-coordinate forms of an equation arising from a uniform asymptotic approximation of high-frequency waves near a caustic (Chap. 5) and of the equation for harmonic fields on the extended projective disc (Chap. 6). Both equations can be put into the form (3.21) in the cartesian $r\theta$-plane, with the x-axis replaced by the line $r = 1$ as a sonic line.

The y-axis divides the domain $\mathscr{D}$ of (3.21) into the subdomains

$$\mathscr{D}^+ = \{(x, y) \in \mathscr{D} | x \geq 0\},$$

and $\mathscr{D}^- = \mathscr{D}\backslash\mathscr{D}^+$. Equation (3.21) is (non-uniformly) elliptic for $(x, y) \in \mathscr{D}^+$.

Theorem 3.1. *Let $u(x, y)$ be a sufficiently smooth solution of (3.21) on $\mathscr{D} \cup \partial\mathscr{D}$, with $\mathscr{K}$ satisfying conditions (3.7) and (3.10). Assume that $\mathscr{K}$ is C^1, and monotonic on $\mathscr{D}^-$. If u vanishes identically on the non-characteristic boundary, then $u \equiv 0$ on all of $\mathscr{D}$.*

Remark. Regarding the required smoothness of the solution u, in the proof we assume as in [40] that u is a twice-continuously differentiable solution of (3.21) up to the boundary, and is identically zero on the noncharacteristic boundary; see also [39]. However, the theorem remains true under the weaker assumption that u is continuous on $\mathscr{D} \cup \partial\mathscr{D}$ and has first partial derivatives which are sufficiently smooth so that the integral

$$I = \int_0^{(x,y)} \left[\mathscr{K}(x)u_x^2 - u_y^2\right]dy - 2u_xu_ydx \tag{3.22}$$

is continuous on $\mathscr{D} \cup \partial\mathscr{D}$; c.f. [38].

Proof. We follow the approach of [38–40]. Differentiating the auxiliary function I under the integral sign, we observe that

$$\frac{\partial}{\partial y}\left(-2u_xu_y\right) - \frac{\partial}{\partial x}\left[\mathscr{K}(x)u_x^2 - u_y^2\right] = \left(-2u_{xy}u_y - 2u_xu_{yy}\right)$$

$$- \left[\mathscr{K}'(x)u_x^2 + 2\mathscr{K}(x)u_xu_{xx} - 2u_yu_{yx}\right]$$

$$= -2u_xu_{yy} - \left[\mathscr{K}'(x)u_x^2 + 2\mathscr{K}(x)u_xu_{xx}\right]$$

$$= -2u_x\left(u_{yy} + \frac{\mathscr{K}'(x)}{2}u_x + \mathscr{K}(x)u_{xx}\right) = 0,$$

using (3.21) and the equivalence of mixed partial derivatives. We conclude that there exists a function $\xi(x, y)$ such that

$$\xi_x = -2u_xu_y \tag{3.23}$$

and

$$\xi_y = \mathcal{K}(x)u_x^2 - u_y^2. \tag{3.24}$$

We will first show that u vanishes identically in $\mathcal{D}^+$. To accomplish this, we must show that u vanishes identically on the sonic line $x = 0$. Once we have shown that, we will have zero boundary conditions on $\mathcal{D}^+$. We will complete the proof for the elliptic region by invoking Proposition 2.1.

Because $u \equiv 0$ on $\mathcal{L}_1$, we conclude that u_x vanishes identically on that horizontal line. Thus we have, by (3.23),

$$\xi_x = 0 \text{ on } \mathcal{L}_1. \tag{3.25}$$

Also, $u \equiv 0$ on $\mathcal{L}_3$, so $u_x = 0$ on that horizontal line as well, implying that

$$\xi_x = 0 \text{ on } \mathcal{L}_3. \tag{3.26}$$

Equations (3.25) and (3.26) imply that on $\mathcal{L}_1$ and $\mathcal{L}_3$, ξ is a function of y only. But y is constant on those two horizontal lines, implying that

$$\xi = c_1 \text{ on } \mathcal{L}_1$$

and

$$\xi = c_2 \text{ on } \mathcal{L}_3,$$

where c_1 and c_2 are constants. On the line $\mathcal{L}_2$, $u \equiv 0$, implying that $u_y = 0$ on that vertical line. Also, $\mathcal{K}(x) > 0$ on $\mathcal{L}_2$. These facts imply, using (3.24), that $\xi_y \geq 0$ on $\mathcal{L}_2$, which in turn implies that

$$c_2 \geq c_1. \tag{3.27}$$

On the line $x = 0$, $\mathcal{K} = 0$, implying by (3.24) that $\xi_y \leq 0$ on that vertical line. This is turn implies that

$$c_2 \leq c_1. \tag{3.28}$$

Inequalities (3.27) and (3.28) are in contradiction unless $c_1 = c_2$. Taking into account that ξ cannot increase with increasing y on the line $x = 0$, it also cannot decrease with increasing y, as it would then have to increase in order to return to its initial value at the endpoint. This implies that $\xi_y = 0$ on the y-axis. Using (3.24) again, we find that on the y-axis,

$$-u_y^2 = 0, \tag{3.29}$$

so the function $u(0, y)$ is constant there. Because

$$u(0, -b) = u(0, b) = 0,$$

that constant is zero. Thus on the rectangle $\partial \mathcal{D}^+$ we have a closed Dirichlet problem having homogeneous boundary conditions.

Proposition 2.1 of Sect. 2.4.1 implies that the C^2 function u attains both its maximum and minimum values on the boundary. Because it is identically zero there, u must be zero in all of $\mathcal{D}^+$.

We obtain the identical vanishing of u on the hyperbolic region by integration along characteristic lines as in [1]. We have

$$d\xi = \xi_x dx + \xi_y dy = \left(-2u_x u_y\right) dx + \left[\mathcal{K}(x)u_x^2 - u_y^2\right] dy.$$

On characteristic lines,

$$dx = \pm\sqrt{-\mathcal{K}(x)}dy$$

and

$$d\xi = \left[\mp 2u_x u_y \sqrt{-\mathcal{K}(x)} + \mathcal{K}(x)u_x^2 - u_y^2\right] dy$$

$$= -\left[\sqrt{-\mathcal{K}(x)}u_x \pm u_y\right]^2 dy \leq 0. \tag{3.30}$$

Thus ξ is nonincreasing in y on any arbitrarily chosen characteristic.

Initially, take $a = 0$.

Because $u \equiv 0$ on the sonic line $x = 0$, we conclude that $u_y = 0$ on that vertical line. So $\xi_x = 0$ on the sonic line by (3.23). Because $K(0) = 0$, we conclude that $\xi_y = 0$ on the sonic line by (3.24) and (3.29). Beginning at the point $(0, -b)$, proceed along Γ_1 to $(m, 0)$ and then along Γ_2 to $(0, b)$. Expression (3.30) implies that ξ will not increase in y along this path from $(0, -b)$ and $(0, b)$. Because ξ is equal to the same constant at those two points, ξ must be constant in y along $\Gamma_1 \cup \Gamma_2$. (If ξ decreased in y at any point along such a path, it would have to increase in y at a later point in order to return to its constant value at $(0, b)$. And it cannot increase in y along a characteristic.) Ascending along the y-axis from the point $(0, -b)$, for any initial point above $(0, -b)$ and any terminal point below $(0, b)$ on the y-axis we can always find a pair of characteristic lines intersecting at some point on the x-axis to the right of $(m, 0)$. We conclude that $\xi_y = 0$ on $\mathcal{D}^-$. But then (3.24) implies that

$$\mathcal{K}(x)u_x^2 = u_y^2 \text{ on } \mathcal{D}^-. \tag{3.31}$$

Because $\mathcal{K} < 0$ on $\mathcal{D}^-$, we are forced to conclude from (3.31) that $u_x = u_y = 0$ on $\mathcal{D}^-$. This is turn implies that u is constant on $\mathcal{D}^-$. Because $u \equiv 0$ on the sonic line, that constant must be zero by the smoothness of u.

Now take $a < 0$. Because $\xi_x = 0$ on $\mathcal{L}_1$ and $\mathcal{L}_3$, ξ remains constant between $(0, -b)$ and $(a, -b)$ and between $(0, b)$ and (a, b). Moreover, ξ_y remains nonpositive along Γ_1 and Γ_2. As we move the initial and terminal points to the right along $\mathcal{L}_1$ and $\mathcal{L}_3$ in $\mathcal{D}^-$, we can always find a pair of characteristic lines which intersect at a point on the x-axis to the right of $(m, 0)$. Arguing as in the case $a = 0$, we again conclude that $u \equiv 0$ on $\mathcal{D}^-$. This completes the proof of Theorem 3.1. $\square$

In the special case in which $\mathcal{K}$ is an analytic function, we do not require a maximum principle, so we do not need to show that $u = 0$ on the line $x = 0$. Rather, we observe that $u_y = 0$ on $\mathcal{L}_2$ as $u \equiv 0$ on that vertical line. Our analysis of the constants c_1 and c_2 implies that $\xi_y = 0$ on $\mathcal{L}_2$ as well. Because in addition, $\mathcal{K} > 0$ on $\mathcal{L}_2$, (3.24) implies that $u_x = 0$ on $\mathcal{L}_2$. We use this last identity as Cauchy data for the Cauchy–Kowalevsky Theorem, to argue that u remains equal to zero as one moves in the negative x-direction away from $\mathcal{L}_2$ along the rectangle $\mathcal{D}^+$. This argument was applied in [40].

An example of a natural type-change function which is *not* analytic is the function

$$\mathcal{K}(x) = \text{sgn}[x], \tag{3.32}$$

which yields an analogue, for equations of Keldysh type, of the Lavrent'ev–Bitsadze equation (2.5). Although such $\mathcal{K}(x)$ is also not C^1, our proof will work for this choice of $\mathcal{K}$ provided (3.21) is suitably interpreted.

Corollary 3.1. *The closed Dirichlet problem for (3.21) on $\mathcal{D}$ is over-determined for u and $\mathcal{D}$ defined as in Theorem 3.1.*

Proof. Suppose that u_1 and u_2 are two such solutions of the open Dirichlet problem for (3.21) on $\mathcal{D}$, with data prescribed only on $\mathcal{L}_1$, $\mathcal{L}_2$, and $\mathcal{L}_3$. Then $U \equiv u_2 - u_1$ satisfies the hypotheses of Theorem 3.1. We conclude that $u_1 = u_2$ in $\mathcal{D}$. That is, any sufficiently smooth solution to (3.21) is uniquely determined by data given on the non-characteristic boundary. So the problem is over-determined for sufficiently smooth solutions if data are given on the entire boundary. This completes the proof. □

Theorem 3.2. *The conclusion of Theorem 3.1 remains true if the Dirichlet conditions on the non-characteristic boundary of $\mathcal{D}$ are replaced by the following mixed Dirichlet–Neumann conditions: $u_y \equiv 0$ on $\mathcal{L}_1$ and $\mathcal{L}_3$; $u \equiv 0$ on $\mathcal{L}_2$.*

Proof. The existence of ξ satisfying (3.23) and (3.24) is established by the same arguments as in the proof of Theorem 3.1. The condition that $u_y = 0$ on $\mathcal{L}_1$ and $\mathcal{L}_3$ implies that $\xi_x = 0$ on those horizontal lines. So the proof of Theorem 3.1 implies that ξ is equal to a constant c_0 on $\mathcal{L}_1$ and $\mathcal{L}_3$. Because $u = 0$ on $\mathcal{L}_2$, (3.23) and (3.24) imply that ξ is equal to c_0 on $\mathcal{L}_2$ as well. The arguments leading to (3.29) imply that u is constant on the line $x = 0$ (but not necessarily equal to zero, as we no longer assume the vanishing of u on the lines $\mathcal{L}_1$ and $\mathcal{L}_3$). So (3.23) and (3.24) imply that ξ is constant on the line $x = 0$. Because of the conditions on $\mathcal{L}_1$ and $\mathcal{L}_3$, that constant is equal to c_0. Thus we conclude that ξ is equal to c_0 on the rectangle $\partial\mathcal{D}^+$.

A direct calculation, using (3.21), (3.23), (3.24), and the identity of mixed partial derivatives, shows that ξ satisfies

$$\mathcal{K}(x)\xi_{xx} + \xi_{yy} + \frac{\mathcal{K}'(x)}{2}\xi_x = 0.$$

Now Proposition 2.1 implies that ξ is a constant (not necessarily zero) in $\mathscr{D}^+$. In particular, (3.23) implies that

$$\xi_x = -2u_x u_y = 0,$$

so $u_x = 0$ and/or $u_y = 0$. If $u_x = 0$, then $u \equiv 0$ in $\mathscr{D}^+$ because $u = 0$ on $\mathscr{L}_2$. If $u_y = 0$, then (3.24) and the constancy of ξ imply that

$$\xi_y = \mathscr{K} u_x^2 = 0.$$

Because $\mathscr{K} > 0$ on $\mathscr{D}^+ \backslash \{x = 0\}$, we conclude that $u_x = 0$ on $\mathscr{D}^+ \backslash \{x = 0\}$. Because $u = 0$ on $\mathscr{L}_2$, we again conclude that $u \equiv 0$ on $\mathscr{D}^+ \backslash \{x = 0\}$. The smoothness of u implies that u is also zero on the line $x = 0$.

The proof that $u \equiv 0$ in $\mathscr{D}^-$ is the same as in the proof of Theorem 3.1.   □

Corollary 3.2. *Let f_1, f_2, and f_3 be given functions defined on the arcs $\mathscr{L}_1$, $\mathscr{L}_2$, and $\mathscr{L}_3$, respectively. The mixed Dirichlet–Neumann problem in which $u_y = f_1$ on $\mathscr{L}_1$, $u_y = f_3$ on $\mathscr{L}_3$, $u = f_2$ on $\mathscr{L}_2$, and any boundary conditions at all are imposed on the characteristic lines, is over-determined for sufficiently smooth solutions of (3.21) in $\mathscr{D}$.*

Regarding the material in this section, see also Theorems 6.1 and 6.2, below, in which an equation of the form (3.21) is defined on domains having geometry which differs from that of the domain of Fig. 3.2. Note that conditions (3.7) and (3.10) are not satisfied in those cases, but the arguments of this section can still be applied.

3.4   A Closed Dirichlet Problem for Distribution Solutions

Although the closed Dirichlet problem is ill-posed for classical solutions, it may be well-posed for solutions having weaker properties.

3.4.1   Almost-Correct *Boundary Conditions*

As we mentioned in Sect. 2.3, the solutions to closed boundary value problems that we will obtain for equations of Keldysh type are a little smoother than generic distribution solutions. The latter fail to lie in a classical function space, whereas our distribution solutions will lie in L^2. They are called *weak solutions* by Berezanskii – c.f. (2.13) of [10], Sect. II.2; but they do not correspond to weak solutions as defined in other standard texts, e.g., [31] or [41].

Define, for a given C^1 function $\mathscr{K}(x, y)$, the space $L^2(\Omega; |\mathscr{K}|)$ and its dual as in Sect. 2.4.2. In this chapter we take the type-change function to depend only on

x, as in Cinquini-Cibrario's original papers. But many of the results will generalize to more complicated choices of $\mathcal{K}$. For example, the results of Sects. 3.3 and 3.4 extend easily to the cold plasma model as presented in Chap. 4, below, in which $\mathcal{K} = x - y^2$; see Sect. 3 of [40] and Sects. 2 and 3 of [43].

Recall that a Hilbert space $\mathcal{H}$ is said to be *rigged* if there is a subspace $\mathcal{V} \subset \mathcal{H}$ for which

$$\mathcal{V} \subset \mathcal{H} \subset \mathcal{V}^*,$$

where $\mathcal{V}^*$ is the space dual to $\mathcal{V}$ in the "test function" topology. If L is formally self-adjoint on distributions, there exists a unique, continuous, self-adjoint extension

$$L : H_0^1(\Omega; \mathcal{K}) \to H^{-1}(\Omega; \mathcal{K}),$$

leading to the rigged triple of Hilbert spaces

$$H_0^1(\Omega; \mathcal{K}) \subset L^2(\Omega) \subset H^{-1}(\Omega; \mathcal{K}).$$

The dual space $H^{-1}(\Omega; K)$ is defined via the negative norm

$$||w||_{H^{-1}(\Omega;K)} = \sup_{0 \neq \varphi \in C_0^\infty(\Omega)} \frac{|\langle w, \varphi \rangle|}{||\varphi||_{H_0^1(\Omega,K)}},$$

Here $\langle \, , \, \rangle$ is the Lax *duality bracket* (or *duality pairing*), motivated by the Schwarz inequality

$$|\langle w, \varphi \rangle| \leq ||w||_{H^{-1}(\Omega;\mathcal{K})} ||\varphi||_{H_0^1(\Omega;\mathcal{K})} \tag{3.33}$$

for $w \in H^{-1}(\Omega; \mathcal{K})$ and $\varphi \in H_0^1(\Omega; \mathcal{K})$; see [9] for a detailed discussion.

The key technical step in the majority of existence proofs in these notes is an *energy inequality* having the form

$$||u||_U \leq C ||Lu||_V, \tag{3.34}$$

where U and V are appropriately chosen spaces of functions or distributions. (In the more general case in which L is not formally self-adjoint, the term Lu on the right-hand side of (3.34) is replaced by $L^* u$; see Appendix A.2.) The choices in our case reduce to:

i) $U = H_0^1(\Omega; \mathcal{K})$, $V = L^2(\Omega)$,

   or

ii) $U = L^2(\Omega; |\mathcal{K}|)$, $V = H^{-1}(\Omega; \mathcal{K})$.

Roughly speaking, choice i) leads to considerably less regularity than choice ii); but (3.34) is considerably harder to establish for choice ii) than for choice i).

In particular, it is sufficient for the existence of an L^2 solution to boundary value problems for the operator equation

$$Lu = f, \tag{3.35}$$

where f is a given distribution depending on (x, y), that inequality (3.34) hold on Ω with U and V defined as in case i). This argument seems to have originated in the work of Berezanskii [8] on so-called "almost-correct" boundary conditions.

In order to prove uniqueness of the solution and the satisfaction of boundary values in a stronger sense than duality, one would like to establish (3.34) for U and V defined according to choice ii). This will be accomplished in Sect. 4.3, for a different choice of type-change function. The failure of the methods of that section to apply in an obvious way to the type-change functions of the present chapter will be analyzed in Sect. 4.5.

3.4.2 The Existence of Distribution Solutions

Consider equations having the form (3.35), where f is a given function or distribution defined on points $(x, y) \in \Omega$. By a *distribution solution* of (3.35) with the boundary condition

$$u(x, y) = 0 \; \forall (x, y) \in \partial\Omega \qquad (3.36)$$

we mean a function $u \in L^2(\Omega)$ such that $\forall \xi \in H_0^1(\Omega; \mathcal{K})$ for which $L^*\xi \in L^2(\Omega)$, we have

$$\left(u, L^*\xi\right) = \langle f, \xi \rangle. \qquad (3.37)$$

Notice that distribution solutions to a homogeneous Dirichlet problem need not vanish on the boundary. (An example of one that does not is based on the fundamental solution to the Tricomi equation: Example 2.4 of [32].)

We will restrict our discussion to the formally self-adjoint operator

$$L_{(\mathcal{K}=x;k=1)}u = x u_{xx} + u_x + u_{yy}, \qquad (3.38)$$

which is a special case of the operator introduced in (2.12).

Lemma 3.1. *Denote by Ω any bounded, connected subdomain of $\mathbf{R}^2$ having piecewise smooth boundary with counter-clockwise orientation. Let u be any C_0^2 function on Ω. Then*

$$\|u\|_{H_0^1(\Omega;x)} \le C \|L_{(x;1)}u\|_{L^2(\Omega)}. \qquad (3.39)$$

Remark. In general the operator on the right is the adjoint of L, which in the case of $L_{(\mathcal{K};1)}$ is L itself. In the proof we will suppress the subscripts of L.

Proof. The proof is similar to the proof of Theorem 2 of [43], and is based on ideas in Sect. 2 of [32].

For a positive constant $\delta \ll 1$, define the function

$$Mu = au + bu_x + cu_y, \qquad (3.40)$$

where $a = -1$, $c = 2(2\delta - 1)\, y$, and

$$b(x) = \begin{cases} \exp(2\delta x/Q_1) & \text{if } x \in \Omega^+ \\ \exp(6\delta x/Q_2) & \text{if } x \in \Omega^- \end{cases}.$$

Here $\Omega^+ = \{x \in \Omega \mid x \geq 0\}$ and $\Omega^- = \Omega \setminus \Omega^+$. Choose $Q_1 = \exp(2\delta\mu_1)$, where $\mu_1 = \max_{x \in \overline{\Omega^+}} x$. Define the negative number μ_2 by $\mu_2 = \min_{x \in \overline{\Omega^-}} x$ and let $Q_2 = \exp(\mu_2)$. For example, if $\Omega = \mathscr{D}$, where $\mathscr{D}$ is the domain of Sect. 3.3, then $\mu_1 = d$ and $\mu_2 = m$. (The constants a and b defined in this section have nothing to do with the constants a and b defined in the preceding section.)

Notice that on Ω^+,

$$2\delta x \leq 2\delta\mu_1 \leq 2\delta\mu_1 e^{2\delta\mu_1} = Q_1 \log Q_1,$$

or

$$\frac{2\delta x}{Q_1} \leq \log Q_1.$$

Exponentiating both sides, we conclude that $b \leq Q_1$ on Ω^+.

Choose $\delta = \delta(\Omega)$ to be sufficiently small so that $6\delta < Q_2$. Then on Ω^-,

$$6\delta x \geq 6\delta\mu_2 = 6\delta \log Q_2 > Q_2 \log Q_2,$$

so $b > Q_2$ on Ω^-.

The coefficient $b(x)$ exceeds zero and is continuous but not differentiable on the y-axis. When we integrate over Ω, it is necessary to introduce a cut along the y-axis, which is analogous to the procedure employed in [32, 33]. The boundary integrals involving a, b, and c on either side of this line will cancel. Integrating by parts using Proposition 2.3 with $\mathscr{K}(x) = x$ and $k = 1$, we obtain

$$(Mu, Lu) = \int\!\!\int_{\Omega^+ \cup \Omega^-} \alpha u_x^2 + \gamma u_y^2 \, dx dy, \tag{3.41}$$

where

$$\alpha_{\Omega^+} = \delta\left[2 - \frac{b}{Q_1}\right] x + \frac{b}{2} \geq \delta x;$$

$$\alpha_{\Omega^-} = \delta\left[2 - 3\frac{b}{Q_2}\right] x + \frac{b}{2} \geq \delta|x|;$$

if δ is sufficiently small, then there is a positive constant ε such that

$$\gamma_{\Omega^+} = 2 + \delta\left(\frac{b}{Q_1} - 2\right) \geq \varepsilon$$

and

$$\gamma_{\Omega^-} = 2 + \delta\left(\frac{3b}{Q_2} - 2\right) \geq \varepsilon.$$

The path integral in Proposition 2.3 does not appear in (3.41) because u has compact support in Ω.

Let

$$\delta' = \min\{\delta, \varepsilon\}.$$

Then

$$\delta' \int\int_\Omega \left(|x|u_x^2 + u_y^2\right) dxdy \leq (Mu, Lu)$$

$$\leq ||Mu||_{L^2}||Lu||_{L^2}$$

$$\leq C(\Omega)\left[\int\int_\Omega \left(|x|u_x^2 + u_y^2\right) dxdy\right]^{1/2} ||Lu||_{L^2(\Omega)},$$

where we have used Proposition 2.2 in obtaining the bound on the L^2-norm of Mu. (In the proof of that proposition it is sufficient for u to be C^1 and to vanish on $\partial\Omega$.) Dividing through by the weighted double integral on the right completes the proof of Lemma 3.1. □

This leads to the following existence result:

Theorem 3.3. *The Dirichlet problem $L_{(\mathscr{K};k)}u = f$ with boundary condition (3.36) possesses a distribution solution $u \in L^2(\Omega)$ for every $f \in H^{-1}(\Omega;\mathscr{K})$ whenever $\mathscr{K} = x$ and $\kappa = 1$.*

Proof. The proof for our case is essentially identical to the proof in the well known case of Tricomi-type operators (c.f. [32], Theorem 2.2). Again we suppress the subscripts of L; but for generality we distinguish L from its adjoint L^*, although under the hypotheses of the theorem the two operators are equal. Define for $\xi \in C_0^\infty$ a linear functional

$$J_f\left(L^*\xi\right) \equiv \langle f, \xi\rangle. \tag{3.42}$$

This functional is bounded on a subspace of L^2 by the inequality

$$|\langle f, \xi\rangle| \leq ||f||_{H^{-1}(\Omega;x)}||\xi||_{H_0^1(\Omega;\mathscr{K})}, \tag{3.43}$$

provided we apply Lemma 3.1 to the variable ξ in the second term on the right. Precisely, J_f is a bounded linear functional on the subspace $\mathscr{M}$ of $L^2(\Omega)$ consisting of elements having the form $L^*\xi$ with $\xi \in C_0^\infty(\Omega)$. The Hahn–Banach Theorem – Theorem 2.1 of Sect. 2.4.4 – allows us to extend J_f to an operator $\mathscr{J}_f$ defined on the closure $\overline{\mathscr{M}}$ of $\mathscr{M}$ in $L^2(\Omega)$, where in the notation of the theorem, $\mathscr{B} = \overline{\mathscr{M}}$, $\ell = J_f$, and $\mathscr{L} = \mathscr{J}_f$. Now we can take ξ to lie in $H_0^1(\Omega;\mathscr{K})$, on the basis of arguments such as those in Sect. 2.4.2, and conclude that $\mathscr{J}_f$ is bounded on $\overline{\mathscr{M}}$

for any $\xi \in H_0^1(\Omega; \mathcal{K}) \ni L\xi \in L^2(\Omega)$. Extending by zero on the orthogonal complement of $\overline{\mathcal{M}}$, we obtain a bounded linear functional on all of $L^2(\Omega)$.

Apply the Riesz Representation Theorem – Theorem 2.2 of Sect. 2.4.4, taking $\mathcal{H} = L^2(\Omega)$, $T = \mathcal{J}_f$, and $X = L^*\xi$. We conclude that there is a vector $u \in L^2$ such that

$$\left(u, L^*\xi\right) = \mathcal{J}_f\left(L^*\xi\right).$$

But the extension of (3.42) to the functional $\mathcal{J}$ implies that

$$\left(u, L^*\xi\right) = \langle f, \xi \rangle$$

$\forall \xi \in H_0^1(\Omega; \mathcal{K}) \ni L\xi \in L^2(\Omega)$. This is our definition (3.37) of distribution solution. $\qquad \square$

These results have been extended to elliptic–hyperbolic operators having a wide range of lower-order terms, which may not be associated with a formally self-adjoint differential operator, but which use the type-change function of Chap. 4; see [43], Sect. 2. Note, however, that the extension of type-change functions of the form $\mathcal{K}(x)$ to type-change functions of the form $\mathcal{K}(x, y)$, whatever their other complications, may simplify the absorption of a nonzero value for the coefficient β in (3.41).

3.5   A Strong Solution to an Open Dirichlet–Neumann Problem

In order to find solutions having stronger regularity properties, we restrict our attention to a favorable class of domains.

3.5.1   D-Star-Shaped Domains

Following Lupo and Payne (Sect. 2 of [34]), we consider a one-parameter family $\psi_\lambda(x, y)$ of inhomogeneous dilations given by

$$\psi_\lambda(x, y) = \left(\lambda^{-\alpha} x, \lambda^{-\beta} y\right)$$

for $\alpha, \beta, \lambda \in \mathbf{R}^+$, and the associated family of operators

$$\Psi_\lambda u = u \circ \psi_\lambda \equiv u_\lambda.$$

Denote by D the vector field

$$Du = \left[\frac{d}{d\lambda} u_\lambda\right]_{|\lambda=1} = -\alpha x \partial_x - \beta y \partial_y. \qquad (3.44)$$

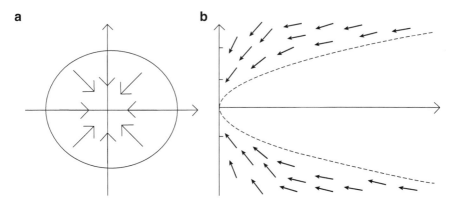

Fig. 3.3 (**a**) The *left-hand domain* is *star-shaped* in the conventional sense; (**b**) the complement of the region enclosed by the parabola in the *right-hand domain* is *star-shaped* with respect to a vector field of the form (3.44) for which $\alpha \neq \beta$

An open set $\Omega \subseteq \mathbf{R}^2$ is said to be *star-shaped* with respect to the flow of D if $\forall \, (x_0, y_0) \in \overline{\Omega}$ and each $t \in [0, \infty]$ we have $F_t \, (x_0, y_0) \subset \overline{\Omega}$, where

$$F_t \, (x_0, y_0) = (x(t), y(t)) = \left(x_0 e^{-\alpha t}, y_0 e^{-\beta t} \right).$$

If a domain is star-shaped with respect to a vector field D, then it is possible to "float" from any point of the domain to the origin along the flow lines of the vector field. If these flow lines are straight lines through the origin ($\alpha = \beta$), then we recover the conventional notion of a star-shaped domain. (See Fig. 3.3.) By an appropriate translation, the origin can be replaced by any point (x_s, y_s) in the plane as a source (or limit point) of the flow. In that case we obtain a translated function $\tilde{F}_t$ for which

$$\lim_{t \to \infty} \tilde{F}_t \, (x_0, y_0) = (x_s, y_s) \quad \forall \, (x_0, y_0) \in \overline{\Omega}.$$

Moreover, whenever a domain is star-shaped with respect to the flow of a vector field satisfying (3.44), the domain boundary will be *starlike* in the sense that

$$(\alpha x, \beta y) \cdot \hat{\mathbf{n}} \, (x, y) \geq 0,$$

where $\hat{\mathbf{n}}$ is the outward-pointing normal vector on $\partial \Omega$; see Lemma 2.2 of [34]. In equivalent notation, given a vector field $V = -(b, c)$ and a boundary arc Γ which is starlike with respect to V, the inequality

$$bn_1 + cn_2 \geq 0 \tag{3.45}$$

is satisfied on Γ.

3.5.2   Admissible Domains

The following theorem is an "admissibility result" in the sense that we assume that the domain Ω supports a host of inequalities with respect to various parameters of the problem and show that solutions exist on such a domain. In that case it is necessary to show that reasonable examples of admissible domains exist, and this will be done subsequently. Note that the term "admissible," when applied to a domain, does not mean exactly the same thing as it does when applied to boundary conditions.

Theorem 3.4 ([44]). *Let Ω be a bounded, connected domain of $\mathbf{R}^2$ having C^2 boundary $\partial\Omega$, oriented in a counterclockwise direction. Let $\partial\Omega_1^+$ be a (possibly empty and not necessarily proper) subset of $\partial\Omega^+$. Let inequality (3.45) be satisfied on $\partial\Omega^+ \backslash \partial\Omega_1^+$. On $\partial\Omega_1^+$ let*

$$bn_1 + cn_2 \leq 0 \tag{3.46}$$

and on $\partial\Omega \backslash \partial\Omega^+$, let

$$-bn_1 + cn_2 \geq 0. \tag{3.47}$$

Let $b(x, y)$ and $c(x, y)$ satisfy

$$b^2 + c^2 \mathcal{K} \neq 0 \tag{3.48}$$

on Ω, with neither b nor c vanishing on Ω^+, and let

$$\mathcal{K} (bn_1 - cn_2)^2 + (c \mathcal{K} n_1 + bn_2)^2 \leq 0 \text{ on } \partial\Omega \backslash \partial\Omega^+. \tag{3.49}$$

Let L be given by (2.13), (2.14) and let EL be symmetric positive, where

$$E = \begin{pmatrix} b & -c\mathcal{K} \\ c & b \end{pmatrix}.$$

Let the Dirichlet condition

$$-u_1 n_2 + u_2 n_1 = 0 \tag{3.50}$$

be satisfied on $\partial\Omega^+ \backslash \partial\Omega_1^+$ and let the Neumann condition

$$\mathcal{K} u_1 n_1 + u_2 n_2 = 0 \tag{3.51}$$

be satisfied on $\partial\Omega_1^+$. Then (2.13), (2.14) possess a strong solution on Ω, in the sense of Sect. 2.5, for every $\mathbf{f} \in L^2(\Omega)$.

Remark. Let the operator L be given by

$$(L\mathbf{u})_1 = xu_{1x} + u_{2y} + \kappa_1 u_1 + \kappa_2 u_2, \tag{3.52}$$

$$(L\mathbf{u})_2 = u_{1y} - u_{2x}, \tag{3.53}$$

where κ_1 and κ_2 are constants. Sufficient conditions for the system to be symmetric positive are

$$2b\kappa_1 - b_x\mathcal{H} - b + c_y\mathcal{H} > 0 \text{ in } \Omega \tag{3.54}$$

and

$$\left(2b\kappa_1 - b_x\mathcal{H} - b + c_y\mathcal{H}\right)\left(2c\kappa_2 + b_x - c_y\right)$$
$$- \left(b\kappa_2 + c\kappa_1 - c_x\mathcal{H} - c - b_y\right)^2 > 0 \text{ in } \Omega. \tag{3.55}$$

Proof. For all points $(\tilde{x}, \tilde{y}) \in \partial\Omega$, decompose the matrix

$$\beta\left(\tilde{x}, \tilde{y}\right) = \begin{pmatrix} \mathcal{H}\left(bn_1 - cn_2\right) & c\mathcal{H}n_1 + bn_2 \\ c\mathcal{H}n_1 + bn_2 & -\left(bn_1 - cn_2\right) \end{pmatrix}$$

into a matrix sum having the form $\beta = \beta_+ + \beta_-$.
   On $\partial\Omega^+\backslash\partial\Omega_1^+$, decompose β into the submatrices

$$\beta_+ = \begin{pmatrix} \mathcal{H}bn_1 & bn_2 \\ \mathcal{H}cn_1 & cn_2 \end{pmatrix}$$

and

$$\beta_- = \begin{pmatrix} -\mathcal{H}cn_2 & \mathcal{H}cn_1 \\ bn_2 & -bn_1 \end{pmatrix}.$$

Then $\beta_-\mathbf{u} = 0$ under boundary condition (3.50). We have

$$\mu^* = (bn_1 + cn_2)\begin{pmatrix} \mathcal{H} & 0 \\ 0 & 1 \end{pmatrix},$$

so condition (3.45) implies that the Dirichlet condition (3.50) is semi-admissible on $\partial\Omega\backslash\partial\Omega_1^+$.
   On $\partial\Omega_1^+$, choose

$$\beta_+ = \begin{pmatrix} -\mathcal{H}cn_2 & \mathcal{H}cn_1 \\ bn_2 & -bn_1 \end{pmatrix}$$

and

$$\beta_- = \begin{pmatrix} \mathcal{H}bn_1 & bn_2 \\ \mathcal{H}cn_1 & cn_2 \end{pmatrix}.$$

Then $\beta_- \mathbf{u} = 0$ under the Neumann boundary condition (3.51), and

$$\mu^* = -(bn_1 + cn_2) \begin{pmatrix} \mathscr{K} & 0 \\ 0 & 1 \end{pmatrix}$$

is positive semi-definite under condition (3.46).

On $\partial\Omega \backslash \partial\Omega^+$, choose $\beta_+ = \beta$ and take β_- to be the zero matrix. Then $\mu = \mu^* = \beta$ and

$$\mu_{11} = \mathscr{K}(bn_1 - cn_2).$$

Because μ_{11} is non-negative by (3.47), μ^* is positive semi-definite by inequality (3.49); so no conditions need be imposed outside the elliptic portion of the boundary. We conclude that the boundary conditions are semi-admissible.

Now we prove admissibility:

On $\partial\Omega^+ \backslash \partial\Omega_1^+$ the null space of β_- is composed of vectors satisfying the Dirichlet condition (3.50), which is imposed on that boundary arc. The null space of β_+ is composed of vectors satisfying the adjoint condition (3.51). On $\partial\Omega_1^+$, this relation is reversed. In order to show that the direct sum of these null spaces spans the two-dimensional space $\mathscr{V}_{|\partial\Omega^+}$, it is sufficient to show that the set

$$\left\{ \begin{pmatrix} 1 \\ n_2/n_1 \end{pmatrix}, \begin{pmatrix} 1 \\ -\mathscr{K} n_1/n_2 \end{pmatrix} \right\}$$

is linearly independent there. Setting

$$c_1 \begin{pmatrix} 1 \\ n_2/n_1 \end{pmatrix} + c_2 \begin{pmatrix} 1 \\ -\mathscr{K} n_1/n_2 \end{pmatrix} = \begin{pmatrix} 0 \\ 0 \end{pmatrix},$$

we find that $c_1 = -c_2$ and

$$-c_2 \left(\frac{n_2^2 + \mathscr{K} n_1^2}{n_1 n_2} \right) = 0. \tag{3.56}$$

Recall that the elliptic boundary is that part of the domain boundary on which the type-change function K is positive. Equation (3.56) can only be satisfied on the elliptic boundary if $c_2 = 0$, implying that $c_1 = 0$. Thus the direct sum of the null spaces of $\beta_\pm$ on $\partial\Omega^+$ is linearly independent and must span $\mathscr{V}$ over that portion of the boundary.

On $\partial\Omega \backslash \partial\Omega^+$, the null space of β_- contains every 2-vector and the null space of β_+ contains only the zero vector; so on that boundary arc, their direct sum spans $\mathscr{V}$.

On $\partial\Omega^+ \backslash \partial\Omega_1^+$, the range $\mathscr{R}_+$ of β_+ is the subset of the range $\mathscr{R}$ of β for which

$$v_2 n_1 - v_1 n_2 = 0 \tag{3.57}$$

for $(v_1, v_2) \in \mathscr{V}$; the range $\mathscr{R}_-$ of β_- is the subset of $\mathscr{R}$ for which

$$\mathscr{K} v_1 n_1 + v_2 n_2 = 0 \tag{3.58}$$

for $(v_1, v_2) \in \mathscr{V}$. Analogous assertions hold on $\partial\Omega_1^+$, in which the ranges of $\mathscr{R}_+$ and $\mathscr{R}_-$ are interchanged. Because if n_1 and n_2 are not simultaneously zero the system (3.57), (3.58) has only the trivial solution $v_2 = v_1 = 0$ on $\partial\Omega^+$, we conclude that $\mathscr{R}_+ \cap \mathscr{R}_- = \{0\}$ on $\partial\Omega^+$.

On $\partial\Omega \setminus \partial\Omega^+$, $\mathscr{R}_- = \{0\}$, so $\mathscr{R}_+ \cap \mathscr{R}_- = \{0\}$ trivially.

The invertibility of E under condition (3.48) completes the proof of Theorem 3.4.

□

By taking $\partial\Omega_1^+$ to be either the empty set or all of $\partial\Omega^+$, Theorem 3.4 implies the existence of strong solutions for either the open Dirichlet problem or the open Neumann problem for the system (2.13), (2.14).

The argument leading to (3.56) suggests that the Tricomi problem (Sect. 3.1) is strongly ill-posed under the hypotheses of the theorem. This is because in the Tricomi problem, data are given on both the elliptic boundary and a characteristic curve; but on characteristic curves, $\mathscr{K}$ satisfies

$$\mathscr{K} = -\frac{n_2^2}{n_1^2}. \tag{3.59}$$

Substituting this equation into (3.56), we find that the equation is satisfied on characteristic curves without requiring the constants c_1 and c_2 to be zero.

However, the theorem is less restrictive if the operator in (2.13) is given by

$$(\mathbf{L}\mathbf{u})_1 = xu_{1x} - u_{2y} + \kappa_1 u_1 + \kappa_2 u_2,$$
$$(\mathbf{L}\mathbf{u})_2 = -u_{1y} + u_{2x}, \tag{3.60}$$

where, again, κ_1 and κ_2 are constants. This variant is analogous to the variant of the Tricomi equation,

$$yu_{xx} - u_{yy} + \text{ zeroth-order terms} = 0,$$

studied in various contexts by Friedrichs [23], Katsanis [28], Sorokina [47,48], and Didenko [20]. In that case, choose

$$E = \begin{pmatrix} b & c\mathscr{K} \\ c & b \end{pmatrix}.$$

Obvious modifications of conditions (3.54) and (3.55) guarantee that the equation

$$EL\mathbf{u} = E\mathbf{f}$$

will be symmetric positive. Condition (3.48) must be replaced by the invertibility condition

$$b^2 - c^2\mathscr{K} \neq 0,$$

which is restrictive on the subdomain Ω^+ rather than on Ω^- as in (3.48). Most importantly, the discussion leading to Table 1 of [29] now applies, with only minor changes, and one can obtain a long list of possible starlike boundaries which result in strong solutions to suitably formulated problems of Dirichlet or Neumann type. In particular, one can formulate a Tricomi problem which is strongly well-posed.

The hypotheses of Theorem 3.4 have a rather formal appearance, but many of the conditions have natural interpretations. For example, inequalities (3.45), (3.46), and (3.47) are satisfied whenever boundary arcs are starlike with respect to an appropriate vector field. Moreover, (3.49) is always satisfied on the characteristic boundary:

Proposition 3.1. *Let Γ be a characteristic curve for (2.13), with the higher-order terms of the operator L satisfying (2.14). Then the left-hand side of inequality (3.49) is identically zero on Γ.*

Proof. We have, using (3.59),

$$
\begin{aligned}
(c\mathcal{K}n_1 + bn_2)^2 &= c^2\mathcal{K}^2 n_1^2 + 2\mathcal{K}cbn_1n_2 + b^2 n_2^2 \\
&= -c^2\mathcal{K}^2 \frac{n_2^2}{\mathcal{K}} + 2\mathcal{K}cbn_1n_2 - b^2\mathcal{K}n_1^2 \\
&= -\mathcal{K}\left(c^2 n_2^2 - 2cbn_1n_2 + b^2 n_1^2\right) \\
&= -\mathcal{K}\left(cn_2 - bn_1\right)^2 .
\end{aligned}
$$

Substituting the extreme right-hand side of this equation into the second term of (3.49) completes the proof.   □

3.5.3 An Example of an Admissible Domain

A simple example which illustrates the hypotheses of Theorem 3.4 can be constructed for the case $\mathcal{K}(x) = x$. In that case the system (2.13), (2.14) can be reduced, by taking $u_1 = u_x$, $u_2 = u_y$, and $\mathbf{f} = 0$, to the original Cinquini-Cibrario equation (3.4). We will consider the formally self-adjoint case, corresponding to the equation

$$
Lu = \mathcal{K}(x)u_{xx} + u_{yy} + \mathcal{K}'(x)u_x,
$$

with $\mathcal{K}(x) = x$ ((3.52) with $\kappa_1 = 1$ and $\kappa_2 = 0$). Choose $b = x + M$, where M is a positive constant which is assumed to be large in comparison with all other parameters of the problem – in particular, $b > 0 \,\forall x \in \overline{\Omega}$; $c = \epsilon(y + 5)$, where ϵ is a small positive constant; $(n_1, n_2) = (dy/ds, -dx/ds)$, where s is arc length on the boundary. We obtain a symmetric-positive system satisfying inequality (3.48) for M sufficiently large.

Fig. 3.4 The domain Ω.
The cusps at the points
$(-4, 0)$ and $(0, \pm 4)$ can be
smoothed out as described in
the text. The *upper curve* is
the characteristic
$y = -2\sqrt{-x} + 4$. The *lower
curve* is the characteristic
$y = 2\sqrt{-x} - 4$

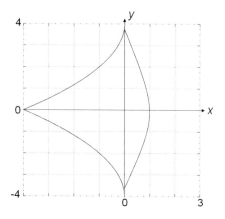

Let the hyperbolic region Ω^- be bounded by intersecting characteristic curves originating on the sonic line. Condition (3.49) is satisfied on $\partial\Omega^-$ by Proposition 3.1. Condition (3.47) is satisfied for M sufficiently large. As a concrete example, let $\partial\Omega^- = \Gamma^- \cup \Gamma^+$, where

$$\Gamma^\pm = \left\{ (x, y) \in \Omega^- | y = \pm 2 \left(\sqrt{-x} - 2 \right) \right\}.$$

These curves intersect at the point $(-4, 0)$. Their intersection is not C^2, but it can be easily "smoothed out" (by the addition of a small noncharacteristic curve connecting the points $(-4 + \delta_0, \pm \delta_1)$ for $0 < \delta_0, \delta_1 \ll 1$) without violating either of the governing inequalities. Let the elliptic boundary $\partial\Omega^+$ be a smooth convex curve, symmetric about the x-axis, with endpoints at $(0, \pm 4)$ on the sonic line. Let the disconnected subset $\partial\Omega_1^+$ of $\partial\Omega^+$ take the form of two small "smoothing curves," on which the slope of the tangent line to $\partial\Omega^+$ changes sign in order to prevent a cusp at the two endpoints. Inequality (3.45) is satisfied on $\partial\Omega^+ \backslash \Omega_1^+$ and, again assuming that M is sufficiently large, inequality (3.46) is satisfied on the two smoothing curves comprising $\partial\Omega_1^+$.

The domain Ω of this construction, shown in Fig. 3.4, is nearly identical to one originally considered by Cinquini-Cibrario — which is illustrated in Fig. 2 on p. 277 of [19]. However, in order for the analysis to apply, the cusps near the points $(-4, 0)$ and $(0, \pm 4)$ should be smoothed out as described above.

Theorem 3.4 implies that strong solutions to an open, homogeneous, mixed Dirichlet–Neumann problem for the first-order inhomogeneous form of Cinquini-Cibrario's equation exist on this natural class of domains for any square-integrable forcing function.

3.6 Relation to Magnetically Dominated Plasmas

Maxwell's equations for a magnetic field **B** (with $\mu_0 = 1$ and ignoring the electric field **E**) assume the form

$$\nabla \times \mathbf{B} = \mathbf{j}, \tag{3.61}$$

$$\nabla \cdot \mathbf{B} = 0. \tag{3.62}$$

The first equation gives the relation between the field and the current $\mathbf{j}$, while the second equation asserts the absence of magnetic monopoles. Equation (3.61) implies that

$$\mathbf{j} \times \mathbf{B} = (\nabla \times \mathbf{B}) \times \mathbf{B}. \tag{3.63}$$

In, e.g., atmospheric plasma or in magnetized thermonuclear fusion in toroidal geometries, the left-hand-side of (3.63) is equal to ∇p, where p is atmospheric pressure in the former case, and the kinetic pressure of a magnetized confined plasma in the latter case. One then obtains the full *magnetostatic equations*, in which the equation

$$\mathbf{j} \times \mathbf{B} = \nabla p \tag{3.64}$$

is appended to (3.61) and (3.62).

In the special case of zero pressure, we obtain the *Beltrami equations*, which consist of (3.62) and the additional equation

$$(\nabla \times \mathbf{B}) \times \mathbf{B} = 0. \tag{3.65}$$

These describe so-called *force-free fields*. Examples include equilibrium magnetic fields in the solar corona.

In the solar physics literature, one usually encounters (3.65) in the form of an eigenvalue equation for the curl operator,

$$\nabla \times \mathbf{B} = \alpha \mathbf{B}, \tag{3.66}$$

where α is a scalar function. Taking into account (3.61), we observe that α has a physical interpretation as the proportionality factor between the magnetic field $\mathbf{B}$ and the current density $\mathbf{j}$ of the medium. In models for the equilibrium magnetic field of the solar corona, α is often taken to be constant or to depend on $\mathbf{r}$. But in general, α varies with position within the magnetic field. Note that any solution of (3.66) automatically satisfies (3.65) by the skew-symmetry of the cross product.

Taking the divergence of both sides in (3.66), one obtains

$$\nabla \cdot (\nabla \times \mathbf{B}) = \nabla \cdot (\alpha \mathbf{B}) = \mathbf{B} \cdot \nabla \alpha, \tag{3.67}$$

using (3.62). But the identity of mixed partial derivatives insures that the left-hand side of (3.67) vanishes for continuously differentiable $\mathbf{B}$, leading to the equation

$$\mathbf{B} \cdot (\nabla \alpha) = 0. \tag{3.68}$$

Equation (3.68) implies that α is constant along the *magnetic lines of force*: curves for which the tangent line at any point lies in the direction of the magnetic field at that point.

If α is known and **B** is unknown, then (3.66) is an elliptic partial differential equation for which Dirichlet or Neumann boundary value problems are expected. If **B** is known and α is unknown, then (3.68) is typically a hyperbolic equation for which the Cauchy problem is expected. Physicists refer to this (somewhat misleadingly) as a *mixed elliptic–hyperbolic structure*; see, e.g., Sect. 2.1 of [3] or (2.1), (2.2) of [5].

In applications to models of the solar corona, one would like to solve (3.66) with condition (3.62) subject to boundary conditions derived from spectro-polarimetric measurements at lower altitudes. In addition to the considerable practical difficulties associated with such measurements (see, e.g., [53]), there is concern that the resulting boundary conditions might not form a well-posed boundary value problem.

Classically, models of the solar corona based on (3.66) and condition (3.62) have been accompanied by Neumann-like boundary conditions having the form [25]

$$B_{n|\partial\Omega} = g,$$

$$\alpha_{|\partial\Omega^+} = h,$$

where g and h are sufficiently smooth functions; B_n is the component of **B** normal to the boundary $\partial\Omega$; and $\partial\Omega^+$ is the part of $\partial\Omega$ on which B_n has a fixed sign – say, the part of the boundary on which B_n exceeds zero. The latter unusual condition corresponds to the prescription of α over one polarity of the magnetic field.

Recently [3], a Dirichlet-type problem was posed for a model of the solar corona based on (3.66) with condition (3.62). In that case one prescribes each of the three cartesian components of the magnetic field **B** on the domain boundary $\partial\Omega$:

$$B_{i|\Omega} = f_i, \quad i = 1, 2, 3,$$

where f_i are three regular functions.

In either case, physicists like to prescribe the asymptotic condition

$$\lim_{|\mathbf{r}|\to\infty} |\mathbf{B}| = 0, \tag{3.69}$$

in order that the solution remain physically meaningful as the outer boundary of the domain is extended out to infinity.

Equations (3.66), (3.62) for axially symmetric fields also arise in plasma models of ball lightning in the Earth's atmosphere [27,50] and of certain extra-galactic jets [51]. But those cases differ in details of the domain geometry, and also with respect to conditions on the fluid pressure, which we have been taking to be zero but which is not zero in the terrestrial atmosphere. In all cases, Neumann-like and/or Dirichlet-like boundary conditions are physically meaningful, but are manifestly well-posed only in the case of α equal to a constant (in which case (3.62) and (3.66) are a first-order variant of the Helmholtz equation). Selections from the vast mathematical literature on physical applications of Beltrami fields include [2, 4, 13, 14, 21, 26].

3.6.1 The Axisymmetric Case

Because force-free magnetic fields in models of the solar corona are axially symmetric, the magnetic field $\mathbf{B}$ can always be written in the form

$$\mathbf{B}(r, \theta) = B_r \hat{\mathbf{e}}_r + B_\theta \hat{\mathbf{e}}_\theta + B_\varphi \hat{\mathbf{e}}_\varphi, \tag{3.70}$$

where $\hat{\mathbf{e}}_r$, $\hat{\mathbf{e}}_\theta$, and $\hat{\mathbf{e}}_\varphi$ are an orthonormal basis for spherical coordinates; in this context, the subscripts r, θ and φ denote components of $\mathbf{B}$ in spherical coordinates (*not* partial differentiation in the direction of those variables); θ denotes the polar spherical coordinate and φ the equatorial spherical coordinate; and

$$B_r = \frac{1}{r^2 \sin\theta} \frac{\partial}{\partial\theta} \left[P(r, \theta) \right], \tag{3.71}$$

$$B_\theta = -\frac{1}{r \sin\theta} \frac{\partial}{\partial r} \left[P(r, \theta) \right], \tag{3.72}$$

$$B_\varphi = \frac{1}{r \sin\theta} Q(r, \theta), \tag{3.73}$$

for scalar functions $P(r, \theta)$ and $Q(r, \theta)$.

Notice that

$$\begin{aligned}
\nabla \cdot \mathbf{B} &\equiv \left(\hat{\mathbf{e}}_r \frac{\partial}{\partial r} + \frac{\hat{\mathbf{e}}_\theta}{r} \frac{\partial}{\partial\theta} + \frac{\hat{\mathbf{e}}_\varphi}{r \sin\theta} \frac{\partial}{\partial\varphi} \right) \cdot \left(B_r \hat{\mathbf{e}}_r + B_\theta \hat{\mathbf{e}}_\theta + B_\varphi \hat{\mathbf{e}}_\varphi \right) \\
&= \frac{1}{r^2} \frac{\partial}{\partial r} \left(\frac{1}{\sin\theta} \frac{\partial P}{\partial\theta} \right) - \frac{1}{r \sin\theta} \frac{\partial}{\partial\theta} \left(\frac{1}{r} \frac{\partial P}{\partial r} \right) = 0,
\end{aligned}$$

so (3.62) is satisfied automatically. Moreover,

$$\begin{aligned}
\nabla \times \mathbf{B} = {}& \frac{1}{r^2 \sin\theta} \left(\frac{\partial Q}{\partial\theta} \right) \hat{\mathbf{e}}_r - \frac{1}{r \sin\theta} \left(\frac{\partial Q}{\partial r} \right) \hat{\mathbf{e}}_\theta \\
&- \left\{ \frac{1}{r \sin\theta} \left(\frac{\partial^2 P}{\partial r^2} \right) + \frac{1}{r^3} \frac{\partial}{\partial\theta} \left[\frac{1}{\sin\theta} \left(\frac{\partial P}{\partial\theta} \right) \right] \right\} \hat{\mathbf{e}}_\varphi. \tag{3.74}
\end{aligned}$$

Thus in order for $\mathbf{B}$ to satisfy (3.66), we require that:

$$\frac{\partial Q}{\partial\theta} = \alpha \frac{\partial P}{\partial\theta} \tag{3.75}$$

in order for the coefficients of $\hat{\mathbf{e}}_r$ in (3.74) to equal those in (3.71);

$$\frac{\partial Q}{\partial r} = \alpha \frac{\partial P}{\partial r}, \tag{3.76}$$

in order for the coefficients of $\hat{\mathbf{e}}_\theta$ in (3.74) to equal those in (3.72). Conditions (3.75) and (3.76) can be satisfied by choosing Q to be a function of P for which

$$\frac{dQ}{dP} = \alpha. \tag{3.77}$$

In order for the components of $\hat{\mathbf{e}}_\varphi$ in (3.74) to equal those in (3.73) we require, taking into account (3.77), that

$$r^2 \frac{\partial^2 P}{\partial r^2} + \frac{\partial^2 P}{\partial \theta^2} - \cot\theta \frac{\partial P}{\partial \theta} + r^2 \alpha Q = 0. \tag{3.78}$$

Thus in the axisymmetric case, taking α to be independent of **B**, the system (3.62), (3.66) reduces to the scalar equation (3.78).

3.6.2   An Open Dirichlet Problem for Cinquini-Cibrario's Equation Which Is Classically Well-Posed

Equation (3.78) in the special case $\alpha = 0$ can be derived from Cinquini-Cibrario's equation (3.4) in the half-plane $x \geq 0$ by the coordinate transformation

$$x = \frac{r^2 \sin^2 \theta}{4}, \quad y = r \cos\theta, \tag{3.79}$$

or equivalently,

$$r = \sqrt{4x + y^2}, \quad \theta = \arctan \frac{2\sqrt{x}}{y}. \tag{3.80}$$

Realizing that, up to a constant, $Q = \alpha P$ by (3.77), we will eventually find that the limitation $\alpha = 0$ is relatively unimportant.

It is possible to solve (3.78) by separation of variables and then paste the solution, which will be defined only within the upper half-plane, onto a particular solution of (3.4) in the lower half-plane.

Proceeding as in Sect. 3 of [19] for the remainder of this section, we seek a solution in the form

$$P(r, \theta) = R(r)T(\theta). \tag{3.81}$$

Substituting (3.81) into (3.78), we obtain

$$r^2 R'' T + R T'' - (\cot \theta) R T'' = 0,$$

implying that

$$\frac{r^2 R''}{R} = \frac{T' \cot \theta - T''}{T} = k,$$

where k is an arbitrary constant, which we will write in the form

$$k = \lambda (\lambda - 1).$$

We obtain the two ordinary differential equations

$$r^2 R'' - \lambda (\lambda - 1) R = 0, \tag{3.82}$$

$$T'' - (\cot \theta) T' + \lambda (\lambda - 1) T = 0. \tag{3.83}$$

Equation (3.82) admits particular solutions of the form $R = r^\lambda$ (and also of the form $R = r^{1-\lambda}$, although solutions of the latter form will not be used). Substituting $\tau = \cos \theta$ into (3.83), that equation assumes the form

$$\left(1 - \tau^2\right) \frac{d^2 T}{d\tau^2} + \lambda (\lambda - 1) T = 0. \tag{3.84}$$

If we write

$$T(\tau) = \left(1 - \tau^2\right) S(\tau),$$

then $S(\tau)$ satisfies the equation

$$\left(1 - \tau^2\right) \frac{d^2 S}{d\tau^2} - 4\tau \frac{dS}{d\tau} + (\lambda - 2)(\lambda + 1) S = 0. \tag{3.85}$$

Choosing λ to be an integer, we find that solutions to (3.85) are well known in the literature on special functions. For example, (3.85) reduces to (13.03) in Chap. 5 of Olver's standard reference [42], taking the parameters μ and ν of that work to equal 1 and $\lambda - 1$, respectively. (Or, in the display equation immediately preceding (15.09) in the same reference, take $\mu = -1$ and $\nu = \lambda - 1$.) The solution can be written in terms of Legendre functions, which we denote by $C_{\lambda-2}^{3/2}(\tau)$. In terms of these functions, we can write a particular solution to (3.84) in the form

$$T(\tau) = \left(1 - \tau^2\right) C_{\lambda-2}^{3/2}(\tau) = \left(\sin^2 \theta\right) C_{\lambda-2}^{3/2} (\cos \theta).$$

Setting $n = \lambda - 2$, we obtain a particular solution to (3.78) in the case $\alpha = 0$:

$$P_n = r^{n+2} \left(\sin^2 \theta\right) C_n^{3/2} (\cos \theta), \tag{3.86}$$

where $C_n^{3/2}$ has the explicit representation

$$C_n^{3/2}(\cos\theta) = \frac{2}{\sqrt{n}} \sum_{s=0}^{m} \frac{(-1)^s \, \Gamma\left(\frac{3}{2} + n - s\right)}{s!\,(n-2s)!} 2^{n-2s} (\cos\theta)^{n-2s} \, ,$$

for $m = n/2$ or $m = (n-1)/2$, depending on whether n is even or odd.

The analysis is only slightly different in the case $Q = \alpha P$, $\alpha \neq 0$, as the zeroth-order term only enters into the simpler of the two equations, (3.82). We obtain, in place of (3.82), the equation

$$r^2 R'' + \left[\alpha^2 r^2 - \lambda\,(\lambda - 1)\right] R = 0, \tag{3.87}$$

which can also be solved in terms of special functions, for example,

$$R(r) = (\alpha r)^{1/2} \, \mathscr{J}_{n+1/2}(\alpha r) \tag{3.88}$$

and

$$T(\tau) = -n\,(n+1) \int_1^\tau \mathscr{P}_n(x)dx. \tag{3.89}$$

In (3.88) and (3.89): $n = -\lambda$, $\mathscr{J}$ is now a standard spherical Bessel function, and $\mathscr{P}$ is a Legendre polynomial – c.f. (18a) and (18b) of [50]; (19b) and (19c') of [51]; and Chap. 5, Sect. 13.1, of [42].

In the other half-plane, $x \leq 0$, a particular solution to (3.4) is given by

$$u_n\,(x, y) = \frac{4}{\pi}\,(n+1)\,(n+2)\,x \int_0^1 t^{1/2}\,(1-t)^{1/2}\,[Y(t)]^n \, dt, \tag{3.90}$$

where

$$Y(t) = y - 2\,(-x)^{1/2} + 4\,(-x)^{1/2}\,t.$$

In terms of x and y, the solution in the half-plane $x \geq 0$ can be written in the form

$$u_n\,(x, y) = \frac{2x}{\sqrt{\pi}} \sum_{s=0}^{m} \frac{(-1)^s \, \Gamma\left(\frac{3}{2} + n - s\right)}{s!\,(n-2-s)!} 2^{n-2-s} \left(4x + y^2\right)^s y^{n-2s}. \tag{3.91}$$

It has been shown by classical arguments ([19], Theorem 9) that the existence of solutions (3.90) and (3.91) in the two half-planes can be used to construct a solution to the following open Dirichlet problem:

Let the domain be given by the region shown in Fig. 3.4, with the boundary arc in the half-plane $x \geq 0$ given explicitly by the curve

$$4x + y^2 = 1. \tag{3.92}$$

(The same arguments will work for boundary arcs of the slightly more general form $4x + (y - y_0)^2 = k$, where k is a positive constant and y_0 is a constant.) It can be shown that a unique solution to (3.4) exists which is equal on the curve C given by (3.92) to a function of the form $x\varphi(y)$, where $\varphi(y)$ is finite and continuous on C, including the endpoints, and vanishes on the y-axis. The solution is analytic in the interior of the domain.

In summary, Cinquini-Cibrario's equation in a half-plane, with a zeroth-order perturbation, is related by a coordinate transformation to an equation for axisymmetric magnetically dominated plasmas. Moreover, solutions in the two half-planes can be pasted together to provide a solution to an open Dirichlet problem for the fully elliptic–hyperbolic equation (3.4).

Boundary value problems for (3.4) restricted to the half plane $x \geq 0$ were studied by Keldysh [30] and have applications to high-speed flow, particularly in certain nonlinear generalizations; see, e.g., [15].

3.7 The Fundamental Solution

The topic of fundamental solutions has potential applications to boundary value problems of all kinds. In particular, it is the main ingredient for solving the Dirichlet problem by the Green's function method. This brief discussion follows the work of S-X. Chen [16].

Recall that a distribution $\mathscr{E}$ is a *fundamental solution* of the operator equation (3.35), for a given operator L and function f, if

$$L\mathscr{E} = \delta. \tag{3.93}$$

Here δ is the *Dirac distribution,* which can be defined as the singular measure (with respect to Lebesgue integration) for which

$$\int_{\mathbf{R}^n} f(x)\delta\,(dx) = f(0), \ x \in \mathbf{R}^n,$$

for all compactly supported continuous functions f; see, e.g., Sect. I.8, Examples 2 and 3, of [54]. The Dirac distribution arises naturally as an element of the negative Sobolev space H^{-s} for $s > n/2$; see Sect. 3.1 of [9] or Example 7.43 of [45] for details of this interpretation.

If $\mathscr{E}$ is a fundamental solution for the operator L, then $\mathscr{E} * f$ is a solution for (3.35), where in this context $*$ denotes the *convolution operator*: for $x, y \in \mathbf{R}^n$,

$$\mathscr{E} * f \equiv \int_{\mathbf{R}^n} \mathscr{E}(y) f\,(x - y)\,dy = \int_{\mathbf{R}^n} \mathscr{E}\,(x - y)\,f(y)dy.$$

Taking $u = \mathscr{E} * f$, we have $\forall\, x \in \mathbf{R}^n$,

$$Lu = L\left(\mathscr{E} * f\right) = L\mathscr{E} * f = \delta * f = f,$$

where we have used (3.93) and the elementary identities $\delta * g = g$ and

$$\partial_i\,(g * h) = (\partial_i g) * h = g * (\partial_i h).$$

The first step in constructing a fundamental solution is often a search for an expression having the right invariance properties. In this case, Cinquini-Cibrario's equation and the Dirac δ-function are both invariant under the coordinate rescaling $s : (x, y) \to (w, z)$, for

$$w = t^2 x, \quad z = ty, \quad t > 0. \tag{3.94}$$

Thus we expect to be able to find *similarity solutions* having the form

$$\varphi = y^\nu F\left(A\frac{x}{y^2} + B\right), \tag{3.95}$$

where A and B are constants to be chosen. Let

$$\mu = A\frac{x}{y^2} + B.$$

Then

$$\varphi_x = Ay^{\nu-2} F'(\mu)$$
$$\varphi_{xx} = A^2 y^{\nu-4} F''(\mu),$$
$$\varphi_y = \nu y^{\nu-1} F(\mu) - 2Axy^{\nu-3} F'(\mu),$$

and

$$\varphi_{yy} = \nu\,(\nu - 1)\,y^{\nu-2} F(\mu)$$
$$-2Ax\,(2\nu - 3)\,y^{\nu-4} F'(\mu) + 4A^2 x^2 y^{\nu-6} F''(\mu).$$

We obtain

$$\mathscr{K}\,(x, y)\,\varphi_{xx} + \varphi_{yy} = y^{\nu-2}\left\{ A^2\left[\frac{\mathscr{K}\,(x, y)}{y^2} + 4\frac{x^2}{y^4}\right] F''(\mu)\right.$$
$$\left. -2A\frac{x}{y^2}\,(2\nu - 3)\,F'(\mu) + \nu\,(\nu - 1)\,F(\mu)\right\}.$$

$$\tag{3.96}$$

Taking $\mathcal{K}(x, y) = x$, then

$$\frac{\mathcal{K}}{y^2} = \frac{x}{y^2} = \frac{\mu - B}{A}.$$

Taking $A = 4$, $B = 1$ in (3.96) allows us to write the equation

$$x\varphi_{xx} + \varphi_{yy} = 0$$

as a particular ordinary differential equation, obtaining

$$y^{\nu-2}\left\{4\mu\,(\mu - 1)\,F''(\mu) - 2\,(2\nu - 3)\,(\mu - 1)\,F'(\mu) + \nu\,(\nu - 1)\,F(\mu)\right\} = 0. \tag{3.97}$$

Recall that the *hypergeometric equation* is an expression of the form

$$z\,(1 - z)\,w''(z) + [c - (a + b + 1)\,z]\,w'(z) - ab \cdot w(z) = 0, \tag{3.98}$$

where a, b, c are given parameters. Applying the method of Frobenius, (3.98) is easily reduced to a series

$$\sum_{n=0}^{\infty}\left\{(n + 1)\,(n + c)\,\alpha_{n+1} - \left[n^2 + (a + b)\,n + ab\right]\alpha_n\right\}z^n = 0,$$

having indicial equation

$$\alpha_{n+1} = \frac{(n + a)\,(n + b)}{(n + 1)\,(n + c)}\alpha_n. \tag{3.99}$$

Writing

$$w(z) = \sum_{n=0}^{\infty}\alpha_n z^n,$$

we obtain a *regular solution*, having the form

$$w(z) = \alpha_0 F\,(a; b; c; z) \equiv \alpha_0\left[1 + \frac{ab}{1!c}z + \frac{a\,(a + 1)\,b\,(b + 1)}{2!c\,(c + 1)}z^2 + \cdots\right]. \tag{3.100}$$

A complete solution is provided by all linear combinations of the two independent solutions $F\,(a; b; c; z)$ and $z^{1-c}F\,(a + 1 - c; b + 1 - c; 2 - c; z)$. The resulting series converges, for $z \in \mathbf{R}$, on $-1 < z < 1$. It converges for $z = \pm 1$ provided $c > a + b$. See, e.g., Sect. 2.1 of [35].

If $y \neq 0$, then multiplying (3.97) by $-1/4$ and taking $\nu = -1$ puts the equation in the form (3.98) with $a = 1/2$, $b = 1$, and $c = 5/2$; c.f. Sect. 4.3.2. This equation, derived in [16], has the two linearly independent solutions

$$F(a;b;c;\mu) = F\left(\frac{1}{2};1;\frac{5}{2};\mu\right)$$

and

$$\mu^{1-c} F(a-c+1;b-c+1;2-c;\mu) = \mu^{-3/2} F\left(-1;-\frac{1}{2};-\frac{1}{2};\mu\right).$$

Evaluating (3.100) for this choice of parameters, we find that only the terms in (3.99) corresponding to $n = 0$ and $n = 1$ are nonzero. That is,

$$F\left(-1;-\frac{1}{2};-\frac{1}{2};\mu\right) = 1 + \frac{(-1)(-1/2)}{(-1/2)}\mu = 1 - \mu,$$

and

$$\varphi = y^{\nu} F(\mu) = y^{-1} \mu^{-3/2}(1-\mu) =$$

$$y^{-1}\left(4\frac{x}{y^2}+1\right)^{-3/2}\left(-4\frac{x}{y^2}\right) = -\frac{4x}{(4x+y^2)^{3/2}}. \tag{3.101}$$

To prove that (3.101) is a fundamental solution, one would ordinarily show that

$$\langle L\mathscr{E}, \psi \rangle = \langle \delta, \psi \rangle = \psi(0,0) \tag{3.102}$$

$\forall \psi(x,y) \in C_0^{\infty}(\mathbf{R}^2)$, where in this context angle brackets denote the duality product of Sect. 2.4.5. However, only a (slightly) weaker assertion is true in this case, because the function φ given by (3.101) has a singularity of order $3/2$ on the characteristic line

$$y^2 + 4x = 0. \tag{3.103}$$

Thus by a *fundamental solution* of (3.4) on any domain including the curve (3.103) we will mean only a distribution $\mathscr{E}$ satisfying the expression

$$\lim_{\epsilon \to 0}\left(\langle \mathscr{E}, L^*\psi \rangle_{|y^2+4x>\epsilon^2} + \langle \mathscr{E}, L^*\psi \rangle_{|y^2+4x<-\epsilon^2}\right) = \psi(0,0) \tag{3.104}$$

for all smooth, compactly supported test functions ψ, where L^* is the operator adjoint to L. In the language of Sect. 2.4.5, the *finite part* of the divergent integral describing the distribution $\mathscr{E}$ (or, in other words, the principal value of the distribution, abbreviated *p.v.*) is a fundamental solution in the ordinary sense of (3.102).

Theorem 3.5 (S-X. Chen [16]). *The distribution*

$$\mathscr{E} = \mathscr{C} \times p.v.\frac{4x\,sgn\left(y^2+4x\right)}{|y^2+4x|^{3/2}}$$

is the fundamental solution of (3.4).

Here $\mathscr{C}$ is a constant coefficient that is computed in the course of the proof.

Proof. We only give the idea of the proof, and refer the reader to Sect. 4 of [16] for the remaining details. (Note the difference in notation between (3.4) and (1) of [16].) For

$$Lu = xu_{xx} + u_{yy},$$

the adjoint operator is given by

$$L^*u = (xu)_{xx} + u_{yy} = xu_{xx} + u_{yy} + 2u_x.$$

In that case

$$uLv - vL^*u = u\left(xv_{xx} + v_{yy}\right) - v\left(xu_{xx} + u_{yy} + 2u_x\right)$$
$$= (uxv_x)_x - [v(xu)_x]_x + \left(uv_y\right)_y - \left(vu_y\right)_y. \qquad (3.105)$$

Integrating the above identity over an appropriate domain of $\mathbf{R}^2$, Stokes' Theorem can be applied on the right-hand side in order to express the left-hand side as a boundary integral.

Now we will construct the boundary of the domain in a way that will be convenient for evaluating (3.104), using (3.105).

Introduce the *characteristic coordinates* [6]

$$\ell = y + 2\sqrt{-x}$$

and

$$m = y - 2\sqrt{-x}$$

on the subdomain $x < 0$. Then the curves $\ell = $ constant and $m = $ constant are the characteristic lines for (3.4). Define the curves $A : m = \epsilon$; $A' : m = -\epsilon$; $B : \ell = -\epsilon$; $B' : \ell = \epsilon$; and

$$\gamma : x = \frac{1}{4}\epsilon^2 \sin^2 \theta, \ y = \epsilon \cos \theta, \ 0 \le \theta \le \pi.$$

(Here θ is the polar angle in the sense that a ray lying along $\theta = 0$ lies along the positive y-axis and a ray lying along $\theta = \pi$ lies along the negative y-axis.) Finally, choose R so large that the interior of the circle $C_R : x^2 + y^2 = R^2$ encloses the support of the test function ψ. Denote by D^+ the domain enclosed by the curves C_R, γ, A, and B. Denote by D^- the domain enclosed by the curves C_R, A', and B' (Fig. 3.5).

Taking $u = \psi$ and $v = \mathcal{E}$ in (3.105), we evaluate this equation on each of the boundary curves in D^+ and D^-, taking into account that ψ has support in the interior of C_R.

Fig. 3.5 The domain of
Theorem 3.5

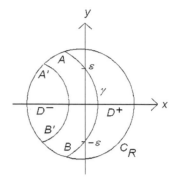

In particular,

$$\int_{D^+} \mathcal{E} L^* \psi \, dxdy = \int_{B \cup \gamma \cup A} \left[x\psi \mathcal{E}_x - \mathcal{E} (x\psi)_y \right] dy + \left(\mathcal{E} \psi_y - \psi \mathcal{E}_y \right) dx.$$

(3.106)

We will completely describe the evaluation of the line integral along the curve γ, and then give a few hints about how the calculation would be done on the remaining curves, referring the reader to [16] for the remainder of the details.

On γ,

$$\mathcal{E} = \frac{4x}{\left(y^2 + 4x\right)^{3/2}} = \frac{\epsilon^2 \sin^2 \theta}{\left(\epsilon^2 \cos^2 \theta + \epsilon^2 \sin^2 \theta\right)^{3/2}} = \frac{\sin^2 \theta}{\epsilon}.$$

$$\mathcal{E}_y = -\frac{3}{2} \frac{4x}{\left(y^2 + 4x\right)^{5/2}} \cdot 2y = -\frac{3 \sin^2 \theta \cos \theta}{\epsilon^2}.$$

$$\mathcal{E}_x = \frac{4}{\left(y^2 + 4x\right)^{3/2}} - \frac{24x}{\left(y^2 + 4x\right)^{5/2}} = \frac{2 \left(2 - 3 \sin^2 \theta\right)}{\epsilon^3}.$$

Moreover,

$$dx = \frac{\epsilon^2}{2} \sin \theta \cos \theta \, d\theta$$

and

$$dy = -\epsilon \sin \theta \, d\theta.$$

Integrating over γ, we write

$$I_\gamma(\epsilon) \equiv I_1 + I_2,$$

where

$$I_1 = \int_\gamma -\mathscr{E}\,(x\psi)_y\,dy + \mathscr{E}\psi_y dx$$

$$= \int_0^\pi \frac{\sin^2\theta}{\epsilon}\left(\psi_x \frac{\epsilon^2 \sin^2\theta}{4} + \psi\right)\epsilon\sin\theta\,d\theta + \int_0^\pi \frac{\sin^2\theta}{\epsilon}\psi_y \frac{\epsilon^2}{2}\sin\theta\cos\theta\,d\theta$$

$$= \int_0^\pi \sin^3\theta\left[\psi_x \frac{\epsilon^2\sin^2\theta}{4} + \psi + \frac{\epsilon}{2}\psi_y\cos\theta\right]d\theta$$

$$= \int_0^\pi \psi\sin^3\theta\,d\theta + O\,(\varepsilon),$$

and

$$I_2 = \int_\gamma \psi x\mathscr{E}_x dy - \psi\mathscr{E}_y dx$$

$$= -\int_0^\pi \psi\left(\frac{1}{4}\varepsilon^2\sin^2\theta\right)\frac{2\,(2-3\sin^2\theta)}{\varepsilon^3}\varepsilon\sin\theta\,d\theta$$

$$+ \int_0^\pi \psi\left(\frac{3\sin^2\theta\cos\theta}{\varepsilon^2}\right)\frac{\varepsilon^2}{2}\sin\theta\cos\theta\,d\theta$$

$$= -\int_0^\pi \psi\sin^3\theta\left[1 - \frac{3}{2}\,(\sin^2\theta + \cos^2\theta)\right]d\theta = \frac{1}{2}\int_0^\pi \psi\sin^3\theta\,d\theta.$$

In the limit as ε tends to zero, γ tends to $(0,0)$ and we obtain

$$I_\gamma(0) = \frac{3}{2}\psi\,(0,0)\int_0^\pi \sin^3\theta\,d\theta = 2\psi\,(0,0). \tag{3.107}$$

Integrating over the curves A, A', B, and B' in characteristic coordinates yields similar estimates. Use the fact that

$$y = \frac{\ell + m}{2}$$

and

$$x = -\left(\frac{\ell - m}{4}\right)^2.$$

Then in characteristic coordinates,

$$\mathscr{E} = \frac{4x}{(y^2 + 4x)^{3/2}}$$

$$= \frac{-4\left[(\ell - m)/4\right]^2}{\left\{\left[(\ell + m)/2\right]^2 - 4\left[(\ell - m)/4\right]^2\right\}^{3/2}} = -\frac{(\ell - m)^2}{4(\ell m)^{3/2}};$$

$$\mathscr{E}_\ell = \frac{1}{2}\frac{(\ell - m)}{(\ell m)^{3/2}}\left[-1 + \frac{3}{4}\frac{(\ell - m)m}{\ell m}\right];$$

$$\mathscr{E}_m = \frac{1}{2}\frac{(\ell - m)}{(\ell m)^{3/2}}\left[1 + \frac{3}{4}\frac{(\ell - m)\ell}{\ell m}\right].$$

Thus

$$\mathscr{E}_y = \mathscr{E}_\ell \ell_y + \mathscr{E}_m m_y = \frac{2(m + \ell)}{8}\frac{(\ell - m)^2}{(\ell m)^{5/2}}.$$

and

$$\mathscr{E}_x = \mathscr{E}_\ell \ell_x + \mathscr{E}_m m_x = \frac{4}{(\ell m)^{3/2}} + \frac{3}{2}\frac{(\ell - m)^2}{(\ell m)^{5/2}}.$$

Substituting these formulas into the right-hand side of (3.106), we obtain a result analogous to (3.107). A similar result can be obtained for the integral over D^-. The sum of the coefficients of $\psi(0,0)$ over the union of the individual boundary arcs is $\mathscr{C}^{-1}$. See Sect. 4 of [16] for details.   □

3.7.1   The Effect of a First-Order Term

As is typical of equations of Keldysh type, the regularity of the solution depends on the magnitude of a lower-order term. To see this, we consider in place of (3.4) the equation

$$L_\kappa u = 0, \tag{3.108}$$

where

$$L_\kappa u = x u_{xx} + \kappa u_x + u_{yy} \tag{3.109}$$

for a constant $\kappa \in [0, 2]$.

Arguing exactly as in (3.96)–(3.101) but including the lower-order term, we obtain in place of (3.101) the expression

$$\varphi = \frac{(-4x)^{1-\kappa}}{(4x + y^2)^{\frac{3}{2}-\kappa}}.$$

Consequently, it is possible to prove the following:

Theorem 3.6 (S-X. Chen [16]). *Let $\kappa > -1/2$ and*

$$\mathcal{E}_\kappa = \begin{cases} 0 & \textit{for } 4x + y^2 \geq 0, \\ \dfrac{(-4x)^{1-\kappa}}{(-4x-y^2)^{\frac{3}{2}-\kappa}} & \textit{for } 4x + y^2 < 0. \end{cases}$$

Then the distribution $T_\kappa = \mathcal{C}_\kappa\, F.P.(E_\kappa)$ is the fundamental solution of (3.108). Here $\mathcal{C}_\kappa$ is a constant depending on κ.

Notice that, not for the first time, the coefficients $\kappa = 1/2$ and $\kappa = 1$ both have special significance, but in this case as "turning points" in the properties of the solution:

At points for which $4x + y^2 < 0$,

$$|4x + y^2| = -\left(4x + y^2\right) \leq -4x.$$

Thus if $\kappa \geq 1$ we have, on the support of E_κ, the identity

$$\frac{(-4x)^{1-\kappa}}{(-4x-y^2)^{\frac{3}{2}-\kappa}} = \frac{(-4x)^{1-\kappa}}{(-4x-y^2)^{1-\kappa}} \cdot \frac{1}{(-4x-y^2)^{\frac{1}{2}}},$$

from which we obtain

$$|E_\kappa| = (-4x)^{1-\kappa}\left(-4x-y^2\right)^{\kappa-1} \cdot \frac{1}{(-4x-y^2)^{\frac{1}{2}}} \leq \frac{1}{(-4x-y^2)^{\frac{1}{2}}}.$$

This greatly simplifies the integrals in the proof. But if $\kappa \leq 1/2$, then the distribution E_κ is singular at points for which $4x + y^2 = 0$, as the associated integrals fail to converge.

References

1. Agmon, S., Nirenberg, L., Protter, M.H.: A maximum principle for a class of hyperbolic equations and applications to equations of mixed elliptic–hyperbolic type. Commun. Pure Appl. Math. **6**, 455–470 (1953)
2. Aly, J.J.: On the uniqueness of the determination of the coronal potential magnetic field from line-of-sight boundary conditions. Solar Phys. **111**, 287–296 (1987)
3. Amari, T., Aly, J.J., Luciani, J.F., Boulmezaoud, T.Z., Mikic, Z.: Reconstructing the solar coronal magnetic field as a force-free magnetic field. Solar Phys. **174**, 129–149 (1997)
4. Amari, T., Boulbe, C., Boulmezaoud, T.Z.: Computing Beltrami fields. SIAM J. Sci. Comput. **31**(5), 3217–3254 (2009)
5. Amari, T., Luciani, J.F., Mikick, Z.: Magnetohydrodynamic models of solar coronal magnetic fields. Plasma Phys. Control. Fusion **41**, A779–A786 (1999)

6. Barros-Neto, J., Gelfand, I.M.: Fundamental solutions for the Tricomi operator. Duke Math. J. **98**, 465–483 (1999)
7. Bateman, H.: Partial Differential Equations. Dover, New York (1944)
8. Berezanskii, Yu.M.: Energy inequalities for some classes of equations of mixed type [in Russian]. Dokl. Akad. Nauk SSSR **132**, 9–12 (1960) [Soviet Math. Doklady **1**, 447–451 (1960)]
9. Berezanskii, Yu.M.: Spaces with negative norms. Russian Math. Surveys **18**(1), 63–95 (1963)
10. Berezanskii, Ju.M.: Expansions in Eigenfunctions of Selfadjoint Operators. American Mathematical Society, Providence (1968)
11. Bers, L.: Mathematical Aspects of Subsonic and Transonic Gas Dynamics. Wiley, New York (1958)
12. Bitsadze, A.V.: Equations of the Mixed Type, Zador, P. (trans). Pergammon, New York (1964)
13. Boulmezaoud, T.A., Amari, T.: On the existence of non-linear force-free fields in three-dimensional domains. Z. Angew. Math. Phys. **51**, 942–967 (2000)
14. Boulmezaoud, T-Z., Maday, Y., Amari, T.: On the linear force-free fields in bounded and unbounded three-dimensional domains. Math. Modelling Numer. Anal. **33**, 359–393 (1999)
15. Čanić, S., Keyfitz, B.: An elliptic problem arising from the unsteady transonic small disturbance equation J. Diff. Equations **125**, 548–574 (1996)
16. Chen, S-X.: The fundamental solution of the Keldysh type operator, Science in China, Ser. A: Mathematics **52**, 1829–1843 (2009)
17. Cibrario, M.: Sulla riduzione a forma canonica delle equazioni lineari alle derivate parziali di secondo ordine di tipo misto. Rendiconti del R. Insituto Lombardo, Ser. II **65**, 889–906 (1932)
18. Cibrario, M.: Alcuni teoremi di esistenza e di unicita per l'equazione $xz_{xx}, +z_{yy} = 0$. Atti R. Acc. Sci. Torino **68** 35–44 (1932–1933)
19. Cibrario, M.: Intorno ad una equazione lineare alle derivate parziali del secondo ordine di tipe misto iperbolico-ellittica. Ann. Sc. Norm. Sup. Pisa, Cl. Sci., Ser. 2 **3**(3, 4), 255–285 (1934)
20. Didenko, V.P.: On the generalized solvability of the Tricomi problem. Ukrain. Math. J. **25**, 10–18 (1973)
21. Flyer, N., Fornberg,B., Thomas, S., Low, B.C.: Magnetic field confinement in the solar corona. I. Force-free magnetic fields. Astrophys. J. **606**, 1210–1222 (2004)
22. Frankl', F.L.: Problems of Chaplygin for mixed sub- and supersonic flows [in Russian]. Izv. Akad. Nauk SSSR, ser. mat. **9**(2), 121–143 (1945)
23. Friedrichs, K.O.: Symmetric positive linear differential equations. Commun. Pure Appl. Math. **11**, 333–418 (1958)
24. Gellerstedt, S.: Quelques problèmes mixtes lour l'équation $y^m z_{xx} + z_{yy} = 0$. Arkiv für Matematik, Astronomi och Fusik **26A**(3), 1–32 (1937)
25. Grad, H., Rubin, H.: Hydromagnetic equilibria and force free fields. In: Proceedings of the 2nd International Conference on Peaceful Uses of Atomic Energy, vol. 31, p. 190. United Nations, Geneva (1958)
26. Hudson, S.R., Hole, M.J., Dewar, R.L.: Eigenvalue problems for Beltrami fields arising in a three-dimensional toroidal magnetohydrodynamic equilibrium problem. Phys. Plasmas **14**, 052505 (2007)
27. Janhunen, P.: Magnetically dominated plasma models of ball lightning. Annales Geophysicae. Atmos. Hydrospheres Space Sci. **9**, 377–380 (1991)
28. Katsanis, T.: Numerical techniques for the solution of symmetric positive linear differential equations. Ph.D. thesis, Case Institute of Technology (1967)
29. Katsanis, T.: Numerical solution of Tricomi equation using theory of symmetric positive differential equations. SIAM J. Numer. Anal. **6**, 236–253 (1969)
30. Keldysh, M.V.: On certain classes of elliptic equations with singularity on the boundary of the domain [in Russian]. Dokl. Akad. Nauk SSSR **77**, 181–183 (1951)
31. Ladyzhenskaya, O.A., Ural'tseva, N.N.: Linear and Quasilinear Elliptic Equations. Academic Press, New York (1968)
32. Lupo, D., Morawetz, C.S., Payne, K.R.: On closed boundary value problems for equations of mixed elliptic-hyperbolic type. Commun. Pure Appl. Math. **60**, 1319–1348 (2007)

33. Lupo, D., Morawetz, C.S., Payne, K.R.: Erratum: On closed boundary value problems for equations of mixed elliptic-hyperbolic type [Commun. Pure Appl. Math. **60**, 1319–1348 (2007)]. Commun. Pure Appl. Math. **61**, 594 (2008)

34. Lupo, D., Payne, K.R.: Critical exponents for semilinear equations of mixed elliptic-hyperbolic and degenerate types. Commun. Pure Appl. Math. **56**, 403–424 (2003)

35. Magnus, W., Oberhettinger, F.: Formulas and Theorems for the Special Functions of Mathematical Physics, J. Werner, trans. Chelsea, New York (1949)

36. Manwell, A.R.: On locally supersonic plane flows with a weak shock wave, J. Math. Mech. **16**, 589–638 (1966)

37. Manwell, A.R.: The Hodograph Equations. Hafner Publishing, New York (1971)

38. Morawetz, C.S.: Note on a maximum principle and a uniqueness theorem for an elliptic-hyperbolic equation. Proc. R. Soc. London, Ser. A **236**, 141–144 (1956)

39. Morawetz, C.S.: Non-existence of transonic flow past a profile. Commun. Pure Appl. Math. **17**, 357–367 (1964)

40. Morawetz, C.S., Stevens, D.C., Weitzner, H.: A numerical experiment on a second-order partial differential equation of mixed type. Commun. Pure Appl. Math. **44**, 1091–1106 (1991)

41. Morrey, C.B.: Multiple Integrals in the Calculus of Variations. Springer, Berlin (1966)

42. Olver, F.W.J.: Asymptotics and Special Functions. A K Peters, Natick (1997)

43. Otway, T.H.: Energy inequalities for a model of wave propagation in cold plasma. Publ. Mat. **52**, 195–234 (2008)

44. Otway, T.H.: Unique solutions to boundary value problems in the cold plasma model. SIAM J. Math. Anal. **42**, 3045–3053 (2010)

45. Renardy, M., Rogers, R.C.: An Introduction to Partial Differential Equations. Springer, Berlin (2004)

46. Riemann, B.: Partielle Differentialgleichungen und deren Anwendung auf Physikalische Fragen. K. Hattendorff, ed. (1861). Nabu Press, 2010

47. Sorokina, N.G.: Strong solvability of the Tricomi problem [in Russian]. Ukrain. Mat. Zh. **18**, 65–77 (1966)

48. Sorokina, N.G.: Strong solvability of the generalized Tricomi problem [in Russian]. Ukrain. Mat. Zh. **24**, 558–561 (1972) [Ukrainian Math. J. **24**, 451–453 (1973)]

49. Tricomi, F.: Sulle equazioni lineari alle derivate parziali di secondo ordine, di tipo misto. Rendiconti Atti dell' Accademia Nazionale dei Lincei Ser. 5 **14**, 134–247 (1923)

50. Tsui, K.H.: A self-similar magnetohydrodynamic model for ball lightnings. Phys. Plasmas **13**, 072102 (2006)

51. Tsui, K.H., Serbeto, A.: Time-dependent magnetohydrodynamic self-similar extragalactic jets. Astrophys. J. **658**, 794–803 (2007)

52. Weitzner, H.: "Wave propagation in a plasma based on the cold plasma model." Courant Inst. Math. Sci. Magneto-Fluid Dynamics Div. Report MF–103, August, 1984

53. Wheatland, M.S.: Reconstruction of nonlinear force-free fields and solar flare prediction. In: Duldig, M. (ed.) Advances in Geosciences, vol. 8: Solar Terrestrial, pp. 123–137. World Scientific, Singapore (2006)

54. Yosida, K.: Functional Analysis. Springer, Berlin (1980)

Chapter 4
The Cold Plasma Model

Because a plasma is a fluid, its evolution must satisfy the equations of fluid dynamics. But because the fluid is composed of electrons and one or more species of ions, the charges on these particles act as sources of an electromagnetic field, which is governed by Maxwell's equations. The presence of this intrinsic field leads to highly nonlinear behavior; and in fact, the dominance of long-range electromagnetic interactions over the short-range interatomic or intermolecular forces is often cited as the defining characteristic of the plasma state. In order to construct a mathematically rigorous model for the plasma which is also accessible to analysis, hypotheses must be imposed which control these nonlinearities. In Sect. 3.6 we assumed that the pressure on the plasma was zero and that magnetic forces dominated over other forces. Those hypotheses reduced the governing equations to the Beltrami equations (3.62), (3.65). In this section we impose a similar physical hypothesis: that the temperature of the plasma is zero.

If the plasma is at zero temperature, then Amontons' Law implies that the pressure term in the equations for fluid motion will also be zero, and the laws of fluid dynamics will enter only through the conservation laws for mass and momentum. Because collisions can be neglected, the fluid aspect of the medium can be ignored as in Sect. 3.6. The cold plasma model approximates the effects of small-amplitude electromagnetic waves, propagating with phase velocities which are sufficiently large in comparison to the thermal velocity of the particles.

Nevertheless, the term *cold plasma* is not used here in the sense in which it appears in the astrophysics literature, in which interstellar plasmas on the order of 10^4 K to 10^5 K are typically referred to as "cold" (see, e.g., [5]). Nor is this model directly applicable to "ultracold" neutral plasmas, having electron temperatures in the range from 1 K to 10^3 K and ion temperatures ranging from 10^{-3} K to 1 K, which recently have been created experimentally [12]. The idealized plasma model considered here is simpler than either of those plasmas, although it remains complex enough to cause serious mathematical difficulties.

T.H. Otway, *The Dirichlet Problem for Elliptic-Hyperbolic Equations of Keldysh Type*,
Lecture Notes in Mathematics 2043, DOI 10.1007/978-3-642-24415-5_4,
© Springer-Verlag Berlin Heidelberg 2012

4.1 Why Study the Cold Plasma Model?

It is plausible that physicists may be interested in the cold plasma model as a toy model for highly rarified plasmas, in which collisions and wall effects are neglected. But why should we be interested in the model in this course?

In a standard ansatz, the field equations of this model are of mixed elliptic–hyperbolic type, and are of Keldysh type on any domain that includes the origin of coordinates. The sonic curve is not a line, as in the Tricomi and Cinquini-Cibrario equations, but a parabola. The Dirichlet problem for this equation is particularly interesting; the closed Dirichlet problem is physically natural but classically ill-posed. The open Dirichlet problem is known to be weakly well-posed [23] but does not appear to be physically natural. So it is reasonable to seek a function space in which the closed Dirichlet problem is also weakly well-posed. This purely mathematical difficulty has been characterized as an "outstanding and significant problem for the cold plasma model" [21].

4.1.1 Very Brief Historical Remarks

The physics of the cold plasma model is grounded in the original work of Tonks and Langmuir in the late 1920s [29], and the model was already well known by the 1950s [1,2,28]. The possibility of singularities arising from plasma inhomogeneities was explored, for the case of electrostatic waves, in the 1970s [14,27], and for the case of electromagnetic waves, in the 1980s [9,30,31]. The analytic aspects of these models have only been investigated in the past two decades [21,23,24,26,32].

4.2 Waves in a Cold Anisotropic Plasma

We initially consider a single particle of mass m and charge $q = Z\delta e$, where $Z \in \mathbf{Z}^+$, $\delta = \pm 1$, and e is the charge on an electron. The particle is subjected only to the Lorentz force

$$\mathbf{F}_L = q\,(\mathbf{E} + \mathbf{v} \times \mathbf{B})\,,$$

where

$$\mathbf{B} = B_0\hat{k}. \tag{4.1}$$

Equation (4.1) implies that the applied magnetic field is *longitudinal* in the sense that its only nonzero component is directed along the positive z-axis.

The equation of motion for the particle is given by Newton's Second Law of Motion:

$$m\frac{d\mathbf{v}}{dt} = \mathbf{F}_L. \tag{4.2}$$

In connection with our fundamental physical hypothesis of zero temperature, we assume that *zero-order quantities* – the plasma density, proportions of ions to electrons, and the background magnetic field – can all be considered static in time and uniform in space, and that *first-order quantities* – the electric field **E** and particle velocities **v** – are expressible as *plane waves*: sinusoidal waves proportional to functions having the form $\exp[i\,(\mathbf{k}\cdot\mathbf{r}-\omega t)]$, where **k** is the propagation vector of the wave and ω is its angular velocity. Thus we write

$$\mathbf{v}\,(x,y,z,t) = \tilde{\mathbf{v}}\,(x,y,z)\exp\left[i\,(\mathbf{k}\cdot\mathbf{r}-\omega t)\right],$$

or

$$\frac{d\mathbf{v}}{dt} = -i\omega\mathbf{v}.$$

Substituting this result into (4.2) yields

$$-im\omega\tilde{\mathbf{v}} = q\left(\tilde{\mathbf{E}}+\tilde{\mathbf{v}}\times\mathbf{B}\right),\tag{4.3}$$

where

$$\mathbf{E}\,(x,y,z,t) = \tilde{\mathbf{E}}\,(x,y,z)\exp\left[i\,(\mathbf{k}\cdot\mathbf{r}-\omega t)\right].\tag{4.4}$$

Initially we will take $\tilde{\mathbf{E}}$ to be a constant vector:

$$\tilde{\mathbf{E}}\,(x,y,z) = E_1\hat{i} + E_2\hat{j} + E_3\hat{k},\tag{4.5}$$

where E_1, E_2, and E_3 are constants, and similarly for $\tilde{\mathbf{v}}$.

Defining the *cyclotron frequency*

$$\Omega = \left|\frac{qB_0}{m}\right|,$$

Equation (4.3) has solutions $\mathbf{v} = (v_1, v_2, v_3)$ satisfying

$$v_1 = \frac{iq}{m\,(\omega^2-\Omega^2)}\,(\omega E_1 + i\delta\Omega E_2);\tag{4.6}$$

$$v_2 = \frac{iq}{m\,(\omega^2-\Omega^2)}\,(\omega E_2 - i\delta\Omega E_1);\tag{4.7}$$

$$v_3 = \frac{iq}{m\omega}E_3.\tag{4.8}$$

Although the above relations were derived for an individual particle, they also hold, in our simplified linear model, for each species of particle in a plasma consisting of electrons and $N - 1$ species of ions. In particular, the plasma current can be written as the sum

$$\mathbf{j} = \sum_{\nu=1}^{N} n_\nu q_\nu \mathbf{v}_\nu,$$

where n_ν is the density of particles having charge magnitude $|q_\nu| = Z_\nu e$.

In the sequel we will only consider the aggregate of particles, in which (4.1)–(4.8) pertain with the quantities $\mathbf{v}$, m, q, Z, δ, and Ω indexed by ν, where $\nu = 1, ..., N$.

4.2.1 Electrostatic Waves

We will call an electric field $\mathbf{E}$ *electrostatic* if it is conservative – that is, if it satisfies the equation

$$\mathbf{E} = -\nabla \Phi,$$

where Φ is the scalar potential [19]. Consider an electrostatic wave in a cold, anisotropic plasma with a two-dimensional inhomogeneity parameterized by two variables, x and z. Then the field potential has the form

$$\Phi (x, y, z) = \varphi (x, z) \exp [i k_2 y],$$

and

$$\mathbf{E} = -\nabla \Phi = (E_1, E_2, E_3) = - \left(\varphi_x e^{i k_2 y}, i k_2 \varphi e^{i k_2 y}, \varphi_z e^{i k_2 y} \right).$$

Introduce the *electric displacement vector*

$$\mathbf{D} = \text{vacuum displacement} + \text{plasma current} = \varepsilon_0 \mathbf{K} \mathbf{E}, \qquad (4.9)$$

where ε_0 is the permittivity of free space,

$$D_i = \varepsilon_0 \sum_{j=1}^{3} K_{ij} E_j, \quad i = 1, 2, 3, \qquad (4.10)$$

and $\mathbf{K} = \left(K_{ij} \right)$ is the *dielectric tensor* (also called the *cold plasma conductivity tensor*). The tensorial nature of this quantity reflects the anisotropy of the plasma.

It can be shown, using (4.6)–(4.9), that

$$\mathbf{K} = \begin{pmatrix} s & -id & 0 \\ id & s & 0 \\ 0 & 0 & p \end{pmatrix}, \qquad (4.11)$$

where s, d, and p are defined in terms of the *plasma frequency*, which for particles of the νth species is given by

$$\Pi_\nu^2 = \frac{n_\nu q^2}{\varepsilon_0 m_\nu},$$

and the *permittivities* R or L of a right- or left-circularly polarized wave travelling in the direction $\hat{k}$; these are given by

$$R = 1 - \sum_{\nu=1}^{N} \frac{\Pi_\nu^2}{\omega^2} \left(\frac{\omega}{\omega + \delta_\nu \Omega_\nu} \right)$$

and

$$L = 1 - \sum_{\nu=1}^{N} \frac{\Pi_\nu^2}{\omega^2} \left(\frac{\omega}{\omega - \delta_\nu \Omega_\nu} \right).$$

Written in terms of these quantities, the entries of the matrix $\mathbf{K}$ are

$$s = \frac{1}{2}(R + L),$$

$$d = \frac{1}{2}(R - L),$$

and

$$p = 1 - \sum_{\nu=1}^{N} \frac{\Pi_\nu^2}{\omega^2}.$$

In terms of $\mathbf{D}$, Maxwell's equations assume the form

$$0 = \nabla \cdot \mathbf{D} = D_{1,x} + D_{2,y} + D_{3,z}. \tag{4.12}$$

We neglect those terms which do not contain derivatives of φ, as φ is assumed to oscillate rapidly. (Regarding the effect of this assumption on the analysis, see the Remark at the end of Sect. 4.3.)

Neither φ nor K_{ij} have any dependence on y, so the problem is effectively two-dimensional. Applying (4.10), the nonzero terms of (4.12) are (setting ε_0 equal to 1)

$$D_{1,x} = -[K_{11}\varphi_{xx} + K_{11,x}\varphi_x + K_{12}\varphi_x ik_2 + K_{13}\varphi_{zx} + K_{13,x}\varphi_z] e^{ik_2 y};$$

$$D_{2,y} = -[K_{21}\varphi_x ik_2 + K_{23}\varphi_z ik_2] e^{ik_2 y};$$

$$D_{3,z} = -[K_{31}\varphi_{xz} + K_{31,z}\varphi_x + K_{32}ik_2\varphi_z + K_{33}\varphi_{zz} + K_{33,z}\varphi_z] e^{ik_2 y}.$$

Collecting terms, we conclude that [27]

$$K_{11}\varphi_{xx} + 2\sigma\varphi_{xz} + K_{33}\varphi_{zz} + \alpha_1\varphi_x + \alpha_2\varphi_z = 0, \tag{4.13}$$

where

$$2\sigma = K_{13} + K_{31};$$

$$\alpha_1 = K_{11,x} + ik_2 (K_{12} + K_{21}) + K_{31,z};$$

$$\alpha_2 = K_{13,x} + ik_2 (K_{23} + K_{32}) + K_{33,z}.$$

Two-dimensional inhomogeneities such as those represented by (4.13) arise in toroidal fields such as those created in tokamaks.

Under our assumptions on $\mathbf{B}_0$, the entries of the matrix K imply that $\sigma = 0$. So we write (4.13) in the form

$$K_{11}\varphi_{xx} + K_{33}\varphi_{zz} + \text{lower-order terms} = 0. \tag{4.14}$$

Equation (4.14) is of either elliptic or hyperbolic type, depending on whether the sign of the product

$$K_{11} \cdot K_{33} = \left(1 - \sum_{\nu=1}^{N} \frac{\Pi_\nu^2}{\omega^2 - \Omega_\nu^2}\right) \cdot \left(1 - \sum_{\nu=1}^{N} \frac{\Pi_\nu^2}{\omega^2}\right) \tag{4.15}$$

is, respectively, positive or negative.

The entry K_{11} changes sign at the *cyclotron* resonances $\omega^2 = \Omega_\nu^2$. The cold plasma model breaks down at these resonances, at which three terms of the dielectric tensor become infinite. The sign of K_{11} also changes at the *hybrid* resonances

$$\sum_{\nu=1}^{N} \frac{\Pi_\nu^2}{\omega^2 - \Omega_\nu^2} = 1. \tag{4.16}$$

In particular, the sign changes at the *lower* hybrid resonance,

$$1 + \frac{\Pi_e^2}{\Omega_e^2} = \frac{\Pi_i^2}{\omega^2}, \tag{4.17}$$

where the subscript e denotes electron frequency, and the subscript i denotes ion frequency. The cold plasma model retains its validity at hybrid resonance frequencies.

The sign of K_{33} changes at the resonance

$$\sum_{\nu=1}^{N} \frac{\Pi_\nu^2}{\omega^2} = 1. \tag{4.18}$$

This is the value at which the frequency of the applied wave equals the plasma frequency of the medium. We may suppose that an electromagnetic wave propagating

Fig. 4.1 Geometry of the
cold plasma model

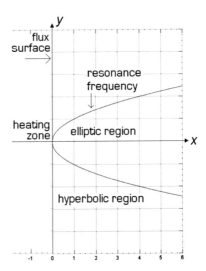

through a plasma does so at a much higher frequency than any of the characteristic
frequencies of the plasma. Otherwise, the plasma magnetic field would oppose the
propagation of the waves (c.f. [13]). Thus we will always take K_{33} to be strictly
positive.

The interesting case is when the coefficient K_{11} models the tangency of a flux
surface to a resonance surface. At the point of tangency, plasma heating might occur
even in the cold plasma model. In two dimensions, flux surfaces (level sets of the
magnetic flux function) can be represented by the lines $x = $ const., and a resonance
surface by a sonic curve having the form

$$x = z^2. \tag{4.19}$$

It has been observed [9, 30] that in such cases a plasma heating zone could lie at the
origin of coordinates (Fig. 4.1).

4.2.2   Electromagnetic Waves

Sonic curves of the form (4.19) seem to arise both in electrostatic plasma waves and
in fully electromagnetic plasma waves. A mathematical model of the latter variety
is necessarily much more complicated than any model for electrostatic waves.
Nevertheless, we include a very brief discussion, based on [30, 31], and Sect. 2.5
of [25].

We adopt for this section the convention that the wave amplitudes are proportional to $\exp[i\omega t]$ in units of c/ω (c.f. 4.4). In that case, Maxwell's equations for
plane waves are

$$\nabla \times \mathbf{E} = -i\,\mathbf{B}, \tag{4.20}$$

$$\nabla \times \mathbf{B} = i\,\mathbf{D} = i\,\mathbf{K}\mathbf{E}. \tag{4.21}$$

Write

$$\mathbf{B} = \nabla \times \mathbf{A} \tag{4.22}$$

and

$$\mathbf{E} = -i\,\mathbf{A} - \nabla\Phi, \tag{4.23}$$

where Φ is a scalar potential and $\mathbf{A}$ is a vector potential.

We also assume that the plasma has axisymmetric geometry, so that we can choose a basis having the form

$$\mathbf{u}_r = \cos\theta\,\hat{i} + \sin\theta\,\hat{j},$$

$$\mathbf{u}_\theta = -\sin\theta\,\hat{i} + \cos\theta\,\hat{j},$$

$$\mathbf{u}_z = -\hat{k},$$

where the subscripts r, θ, and z denote respectively the radial, angular, and axial components of the vector. In this case we have

$$\mathbf{E} = \left(E_r\,(r,z)\,\mathbf{u}_r + E_\theta\,(r,z)\,r\mathbf{u}_\theta + E_z\,(r,z)\,\mathbf{u}_z \right) e^{im\theta},$$

and similarly for $\mathbf{B}$. Letting $m = 0$, the waves preserve the axial symmetry of the underlying static plasma medium, as the wave vector satisfies $\mathbf{k} = (k_r, 0, k_z)$. Then

$$\nabla \times \mathbf{E} = \frac{1}{r}\,[E_{z,\theta}\mathbf{u}_r + E_{r,z}\,(r\mathbf{u}_\theta) + (rE_\theta)_{,r}\,\mathbf{u}_z]$$

$$-\frac{1}{r}\,[E_{z,r}\,(r\mathbf{u}_\theta) + rE_{\theta,z}\mathbf{u}_r + E_{r,\theta}\mathbf{u}_z]$$

$$= -E_{\theta,z}\mathbf{u}_r + (E_{r,z} - E_{z,r})\,\mathbf{u}_\theta + \frac{1}{r}\,(rE_\theta)_{,r}\,\mathbf{u}_z. \tag{4.24}$$

Now (4.20) implies

$$-i\,E_{\theta,z} = B_r \tag{4.25}$$

and

$$\frac{(rE_\theta)_{,r}}{r} = -i\,B_z. \tag{4.26}$$

Equations (4.25) and (4.26) imply, using (4.20), that B_r and B_z can each be expressed in terms of derivatives of E_θ. Similarly, (4.21) implies that the other cylindrical components of $\mathbf{E}$ and $\mathbf{B}$ can be expressed as suitable derivatives of E_θ and B_θ.

Applying (4.20) to the middle term of the last identity in (4.24), we obtain

$$E_{r,z} - E_{z,r} = -i B_\theta. \tag{4.27}$$

Because **E** and **B** are analogous and the left-hand and middle terms of (4.21) are analogous to (4.20) with a change of sign, we have

$$D_r = i B_{\theta,r}$$

and

$$D_z = -i \frac{(r B_\theta)_{,z}}{r}.$$

Now the extreme right-hand side of (4.21) yields E_r and E_z (see (22), (23) of [30]). We obtain

$$B_{r,z} - B_{z,r} = i D_\theta = i \left(K_{\theta r} E_r + K_{\theta\theta} E_\theta + K_{\theta z} E_z \right), \tag{4.28}$$

which completes the system of equations for E_θ and B_θ.

Equations (4.27), (4.28) can be associated to the energy functional

$$\begin{aligned}
\mathcal{E} = \int \{ & \left[\nabla \left(r E_\theta^* \right) \cdot \nabla \left(r E_\theta \right) \right] / r^2 + \left[\nabla \left(r B_\theta^* \right) \cdot \mathbf{K} \nabla \left(r B_\theta \right) \right] / r^2 \Delta \\
& + i E_\theta \left[\left(r B_\theta^* \right)_{,r} \left(K_{zr} K_{r\theta} - K_{z\theta} K_{rr} \right) / r + B_{\theta,z}^* \left(K_{r\theta} K_{zz} - K_{rz} K_{z\theta} \right) \right] / \Delta \\
& - i E_\theta^* \left[\left(r B_\theta \right)_{,r} \left(K_{zr} K_{\theta r} - K_{\theta z} K_{rr} \right) / r + B_{\theta,z} \left(K_{\theta r} K_{zz} - K_{zr} K_{\theta z} \right) \right] / \Delta \\
& - B_\theta^* B_\theta - E_\theta^* E_\theta \left[\det (\mathbf{K}) \right] / \Delta \} r \, dr \, dz, \tag{4.29}
\end{aligned}$$

where

$$\nabla = \frac{\partial}{\partial r} r + \frac{\partial}{\partial z} z,$$

and

$$\Delta = K_{rr} K_{zz} - K_{rz} K_{zr}.$$

$\mathcal{E}$ is self-adjoint provided the dielectric tensor **K** is self-adjoint. Form a right-handed orthogonal set $(\mathbf{v}, \boldsymbol{\theta}, \mathbf{u})$, where

$$\mathbf{u} = \sin \beta r + \cos \beta k$$

and

$$\mathbf{v} = \cos \beta r - \sin \beta k.$$

The basis is chosen so that **u** points in the poloidal direction and **v** is orthogonal to it. The magnetic field vector can be written in the form

$$\mathbf{B} = B_0 \left[\cos \alpha \theta + \sin \alpha \left(\sin \beta r + \cos \beta z \right) \right].$$

In this expansion α, β, and B_0 depend only on r and z. The variational equations of $\mathcal{E}$ now assume the form of a second-order system in which the differential operator for one of the equations is essentially the Laplacian $\mathcal{L}$. The differential operator for the other equation looks like

$$\mathcal{L} + (\mathbf{u} \cdot \nabla)^2. \tag{4.30}$$

That is, the second of the two differential operators in that equation is the only mathematically interesting part of the entire model. In terms of our choice of basis, we have, from (4.30),

$$r\nabla \cdot \left[\left(\frac{\xi}{r^2 \Delta} \right) \nabla (rB_\theta) \right] - r\nabla \cdot \left[\left(\frac{\zeta \sin^2 \alpha}{r^2 \Delta} \right) (\mathbf{u} \cdot \nabla) (rB_\theta) \mathbf{u} \right]$$

$$-i\theta \cdot \nabla (rB_\theta) \times \nabla \left(\frac{\mu \cos \alpha}{r\Delta} \right) + B_\theta$$

$$= (r\Delta)^{-1} \left[\mu (\zeta - \xi) \sin \alpha \mathbf{u} \cdot \nabla (rE_\theta) + i \left(\mu^2 - \xi \zeta \right) \sin \alpha \cos \alpha \mathbf{v} \cdot \nabla (rE_\theta) \right], \tag{4.31}$$

for

$$\xi = 1 + \sum_{\nu=1}^{N} \frac{\Pi_\nu^2}{\Omega_\nu^2 - \omega^2},$$

$$\zeta = \xi + \sum_{\nu=1}^{N} \frac{\Pi_\nu^2}{\omega^2} - 1,$$

and

$$\mu = \sum_{\nu=1}^{N} \frac{\Pi_\nu^2 \Omega_\nu}{\omega \left(\Omega_\nu^2 - \omega^2 \right)}.$$

The mathematical condition that (4.31) is elliptic for negative values of ξ is equivalent to the physical condition for so-called *lower-hybrid* frequencies, at which

$$1 + \frac{\Pi_e^2}{\Omega_e^2} < \frac{\Pi_i^2}{\omega^2},$$

c.f. (4.17). Because ξ is a function of r and z, we can define a new variable $\eta (r, z)$ in which the curves $\eta = $ constant are orthogonal to the curves $\xi = $ constant. Rewriting (4.31) in (ξ, η)-coordinates, we find that the type of the equation changes when

$$\mathbf{u} \cdot \nabla \xi = 0.$$

Physically, this means that (4.31) changes type when flux surfaces coincide with resonance surfaces. The highest-order terms of that equation can be written in the form

$$Lu = f\,(\xi, \eta)\left[\xi u_{\xi\xi} + M\,(\xi, \eta)\,u_{\eta\eta}\right],\tag{4.32}$$

where $u = u\,(\xi, \eta)$ is a scalar function; f and M are given scalar functions which are sufficiently smooth near $\xi = 0$, and $M > 0$.

If the curve representing the flux surface in two dimensions is collinear with the resonance curve as in (4.32), then the plasma can be treated as a perpendicularly stratified medium. In that case (4.31) reduces to a form of Cinquini-Cibrario's equation. If the resonance curve is tangent to the curve representing the flux surface, then we are in a more interesting case. In that case the simplest model for the operator L of (4.31) is an operator having highest-order terms of the form

$$\tilde{L}u = \left(x - y^2\right)u_{xx} + u_{yy}\tag{4.33}$$

(where the switch from z to y represents the necessity of a change of variables with respect to (4.19); c.f. Sect. 2.5.2 of [25]). And that is the equation which we will study in this chapter.

The preceding analysis, which is based on physical reasoning, suggests that the closed Dirichlet problem should be well-posed for a model of electromagnetic wave propagation in zero-temperature plasma. However, the closed Dirichlet problem has been shown to be ill-posed, in the classical sense, for the equation

$$\left(x - y^2\right)u_{xx} + u_{yy} + \frac{1}{2}u_x = 0\tag{4.34}$$

on a typical domain [21]. The proof of ill-posedness is similar to that of Theorem 3.1. This result is problematic for the cold plasma model, as (4.34) is a special case of the class of equations

$$\left(x - y^2\right)u_{xx} + u_{yy} + \text{ lower order terms } = 0\tag{4.35}$$

which, according to our argument, is precisely the kind of equation that should model the behavior of the plasma at a possible heating zone. This leads us to ask whether a well-posed problem with closed boundary data can be formulated for such an equation in a suitably weak sense.

The case in which the lower-order terms of (4.35) are of the form κu_x, where κ is a constant in $[0, 2]$, was considered in [24]. The existence of distribution solutions to the homogeneous closed Dirichlet problem is shown by methods similar to those applied in in Sect. 3.4. In Sect. 4.3 we will find that considerably greater regularity can be derived for solutions in the special case $\kappa = 1$.

The result which may be most fundamental for the mathematics of the cold plasma model is "folklore," in the sense that various special cases are often cited but a proof does not exist in the literature (apart from some technical remarks in the introduction to [21]):

Theorem 4.1. *Consider equations having the form*

$$A \left(x + ay^2\right) u_{xx} + B u_{yy} = 0, \tag{4.36}$$

where A, B, a and κ are nonzero constants which satisfy

$$A = -\frac{2aB}{\kappa}. \tag{4.37}$$

Any such equation is equivalent under the coordinate transformation $(x, y) \rightarrow (\xi, \eta)$, *where*

$$\xi = \kappa A \left(x + ay^2\right), \tag{4.38}$$

to an equation of Tricomi type except at the origin of coordinates in (x, y).

Remarks. If $A = B = 1$, $a = -1$, and $\kappa = 2$, then the assertion of the theorem applies to the equation $\tilde{L}u = 0$, where $\tilde{L}$ is given by (4.33). If $A = 1$, $B = -1$, $a = -1$, and $\kappa = -2$, then the assertion of the theorem applies to (B.6) of Appendix B. If $A = 1$, $B = -1$, and $\kappa = 2a$, then the assertion of the theorem applies to (9) of [27].

Proof. Proceed as in Sect. 3.2 (substituting xy-coordinates for the xz-coordinates of that section). Choose, in the notation of that section,

$$\alpha = A \left(x + ay^2\right) \text{ and } \gamma = B. \tag{4.39}$$

In order for the cross term $u_{\xi\eta}$ to vanish in the new coordinates (ξ, η), we require that

$$0 = \xi\xi_x \eta_x + \gamma\xi_y \eta_y = A\kappa \left[A\kappa \left(x + ay^2\right) \eta_x + 2aBy\eta_y\right],$$

where we have used (4.38). This condition can be satisfied by choosing

$$\eta = A\kappa \left(xy + a\frac{y^3}{3}\right) \tag{4.40}$$

and taking into account (4.37). The Jacobian of the transformation $(x, y) \rightarrow (\xi, \eta)$ will be nonvanishing provided

$$0 \neq \xi_x \eta_y - \xi_y \eta_x = (A\kappa)^2 \left(x - ay^2\right), \tag{4.41}$$

a condition which cannot be satisfied at the origin. Substituting (4.39) and (4.40) into (3.17), we find that, away from the origin, the coordinate change $(x, y) \rightarrow (\xi, \eta)$ defined by (4.38), (4.40) takes (4.36) into the equation

$$u_{\xi\xi} - \frac{\xi}{2a\kappa A} u_{\eta\eta} = 0,$$

an equation of Tricomi type. But in addition to the vanishing at the origin of the Jacobian in (4.41), condition ii) of Sect. 3.2 is satisfied at the origin of coordinates in $(\xi\,(x,y)\,,\eta\,(x,y))$ by (4.38); so the transformed equation cannot be of Tricomi type at the origin. $\qquad\qquad\qquad\qquad\qquad\qquad\qquad\qquad\qquad\qquad\qquad\qquad\qquad\qquad\square$

A corollary of this result is the assertion that lower-order terms will not play an important role in the existence and regularity of solutions except on domains which include the origin of coordinates. Because for physical reasons we will only consider such domains, it will be necessary to prescribe the form of the lower-order terms in the following section.

4.3   A Closed Dirichlet Problem Which Is Weakly Well Posed

Let $\mathscr{K}\,(x,y)=x-y^2$. Following Lupo, Morawetz, and Payne [15, 16], we define a *weak solution* of the equation

$$Lu \equiv [\mathscr{K}\,(x,y)\,u_x]_x + u_{yy} = f\,(x,y) \tag{4.42}$$

on Ω, with boundary condition

$$u(x,y) = 0 \,\forall\,(x,y) \in \partial\Omega, \tag{4.43}$$

to be a function $u \in H^1_0(\Omega;\mathscr{K})$ such that $\forall \xi \in H^1_0(\Omega;\mathscr{K})$ we have

$$\langle Lu, \xi\rangle \equiv -\int\int_\Omega \left(\mathscr{K} u_x \xi_x + u_y \xi_y\right) dx dy = \langle f, \xi\rangle\,,$$

where, as before, $\langle\,,\,\rangle$ represents the duality pairing between $H^1_0\,(\Omega,\mathscr{K})$ and $H^{-1}\,(\Omega,\mathscr{K})$. The function f is assumed to be known. In this case the existence of a weak solution is equivalent to the existence of a sequence $u_n \in C^\infty_0(\Omega)$ such that

$$||u_n - u||_{H^1_0(\Omega;\mathscr{K})} \to 0 \text{ and } ||Lu_n - f||_{H^{-1}(\Omega;\mathscr{K})} \to 0$$

as n tends to infinity.

We employ an integral variant of the *abc* method, which was introduced by Didenko [6] and developed by Lupo and Payne [17]. (In [26], this method is misleadingly identified with the "dual variational method," also developed by Lupo and Payne in [17]. See Problem 18, part i), of Appendix B.1.) Denote by v a solution to the boundary value problem

$$Hv = u \text{ in } \Omega \tag{4.44}$$

for $u \in C^\infty_0(\Omega)$, with

$$Hv = av + bv_x + cv_y, \tag{4.45}$$

Fig. 4.2 Geometry of the
auxiliary problem
(4.44)–(4.45)

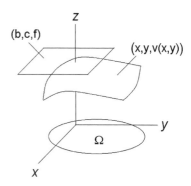

v vanishing on $\partial\Omega \setminus (0,0)$, and

$$\lim_{(x,y)\to(0,0)} v(x,y) = 0. \tag{4.46}$$

Assume that Ω is star-shaped with respect to the flow of the vector field $V = -(b,c)$, in the sense of Sect. 3.5.1; let $b = mx$ and $c = \mu y$, where μ and m are positive constants and a is a negative constant; also assume that the point $(x, y) = (0,0)$ lies on $\partial\Omega$.

Proposition 4.1 (Lupo–Morawetz–Payne [15]). *Under the assumptions of the preceding paragraph, a solution $v \in C^0\left(\overline{\Omega}\right) \cap H_0^1\left(\Omega; \mathcal{K}\right)$ to the boundary value problem (4.44)–(4.46) exists.*

Proof. The proof that we give is somewhat different in its details from the one given in [15] (as Step 1 in the proof of Lemma 3.3 of that reference). The reason is that in [15], the coefficients are only Lipschitz continuous. For our purposes we can take them to be actually smooth, permitting somewhat more explicit calculations.

Proceeding as in the classical method of characteristics, we write (4.44), for H defined as in (4.45), as the dot product of vectors. Defining

$$f(x, y, v(x, y)) \equiv u(x, y) - av(x, y),$$

equations (4.44), (4.45) imply the vector equation

$$(b, c, f) \cdot (v_x, v_y, -1) = 0. \tag{4.47}$$

Denoting by S the solution surface $(x, y, v(x, y))$ for the system (4.44), (4.45), we note that because $(v_x, v_y, -1)$ is the normal vector to S, (4.47) implies that the vector (b, c, f) lies in the tangent plane to S (Fig. 4.2).

Representing the tangent plane to S as the space of all "velocity vectors" on S, we parameterize the points of S by $(x(t), y(t), v(x(t), y(t)))$. Because u has compact support and the solution v must vanish on $\partial\Omega \setminus (0,0)$, we take $f = 0$ there. Thus if (x_0, y_0) is a point on $\partial\Omega \setminus (0,0)$, in order to compute the trajectory of an integral curve of the solution having initial value at that point, we solve the system

$$\frac{dx}{dt} = b = mx, \tag{4.48}$$

$$\frac{dy}{dt} = c = \mu y. \tag{4.49}$$

Integrating equations (4.48), (4.49) directly and applying the initial condition, we obtain

$$(x(t), y(t)) = \left(x_0 e^{mt}, y_0 e^{\mu t}\right) \equiv F_t (x_0, y_0),$$

where F_t is the function defined in Sect. 3.5.1. (In terms of the definition given in Sect. 3.5.1, $\alpha = -m$, $\beta = -\mu$, and $V = (-b, -c)$. The proofs of this section will work as long as m and μ have the same sign and a has the opposite sign.) So the hypothesis that Ω is star-shaped with respect to the flow of $-V$ means that any streamline originating on a point (x_0, y_0) on $\partial\Omega \setminus (0, 0)$ will remain in that domain, possibly becoming singular at its terminal point at the origin (Fig. 4.3).

In order to determine whether the flow of the solution will become singular at the origin, we explicitly compute the streamlines. Solving the equation $x = x_0 \exp[mt]$ for t, we obtain $t = m^{-1} \ln (x/x_0)$. Substituting this value into the equation $y = y_0 \exp[\mu t]$, we obtain

$$y = k_1 x^{\mu/m}, \tag{4.50}$$

where

$$k_1 = \frac{y_0}{x_0^{\mu/m}}.$$

Notice that if we compute the characteristic lines for the equation (4.44), (4.45) in the usual way, by integrating the differential equation

$$bdy = cdx,$$

we obtain the equation

$$y = e^C x^{\mu/m} \tag{4.51}$$

for the characteristic lines, where e^C is an integration constant, which is of course a subset of the family (4.50). Because these curves are real on the half-plane $x \geq 0$, we conclude that (4.44), (4.45) is hyperbolic on the domain Ω. Prescribing homogeneous data on the curve $\partial\Omega \setminus (0, 0)$ corresponds to a Cauchy problem for this hyperbolic equation, which we expect to be well-posed.

However, adding the condition (4.46) to the other conditions converts the homogeneous Cauchy problem into a (possibly singular) homogeneous Dirichlet problem, which is *not* generally well-posed for hyperbolic equations.

Because the origin lies on the boundary and u has compact support, we can restrict our attention to an ε-neighborhood N_ε of the boundary,

Fig. 4.3 A collection of
streamlines of V for the case
$\mu = 1$, $m = 4$. The *hatched
line* is the sonic curve
$y = \sqrt{x}$ for the cold plasma
model equation

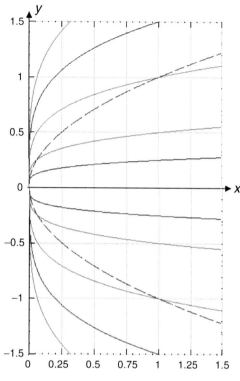

Fig. 4.4 The construction
of N_ε

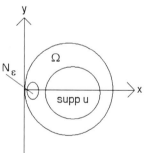

$$N_\varepsilon\left(\partial\Omega\right) = \left\{(x, y) \in \overline{\Omega}\,\middle|\,\mathrm{dist}\left((x, y), \partial\Omega\right) \leq \varepsilon\right\},$$

where ε is so small that N_ε lies in $\overline{\Omega}$ but outside the support of u in Ω (Fig. 4.4).
In this subdomain we re-initialize, and solve the Cauchy problem for the homogeneous equation

$$mxv_x + \mu yv_y = |a|v. \tag{4.52}$$

It is easy to check that a general solution in N_ε has the form

Fig. 4.5 The construction of $\tilde{N}_{\varepsilon}$

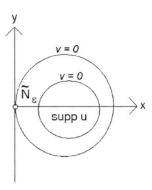

$$v(x, y) = \varphi \left(\frac{x^{\mu}}{y^m} \right) y^{|a|/\mu}, \tag{4.53}$$

where φ is an arbitrary C^1 function that may be prescribed along a non-characteristic curve in N_{ε}. Following the evolution of the solution (4.53) along the characteristic lines (4.51), we obtain (4.46) as in (3.13) of [15].   □

Remarks. Define $\tilde{N}_{\varepsilon}$ to be the set $\Omega \setminus \mathrm{supp}(u)$ – that is, the complement in Ω of the support of u; see Fig. 4.5. Observe that (4.44), (4.45) reduce in that entire region to (4.52) (not just in the smaller region N_{ε}). Moreover, $v \equiv 0$ on $\partial \tilde{N}_{\varepsilon} \setminus \{(0,0)\}$. An obvious solution to this equation is the trivial solution $v \equiv 0$ in $\tilde{N}_{\varepsilon}$. The boundary condition $v \equiv 0$ on $\partial \Omega \setminus \{(0,0)\}$ can be re-initialized on the boundary $\partial (\mathrm{supp}(u)) \setminus \{(0,0)\}$ of the subdomain on which (4.44), (4.45) are *not* equivalent to (4.52). Thus we can assume that the solution v guaranteed by Proposition 4.1 vanishes identically in a neighborhood of the punctured boundary $\partial \Omega \setminus \{(0,0)\}$. (This assumption will be used in evaluating the boundary integral of (4.56), below.)

The identical vanishing of v in a neighborhood of $\partial \Omega \setminus \{(0,0)\}$ excludes the possibility of a blow-up in the derivatives of (4.53) as the punctured boundary is approached. Because v is C^1 in the interior, v has bounded variation along any initial curve. We conclude that v lies in $H_{loc}^1 (\Omega; \mathscr{K})$. In conjunction with its boundary behavior, this implies that in fact $v \in H_0^1 (\Omega; \mathscr{K})$. (See the discussion of [8], Sect. 7.5, which extends immediately to weighted spaces.)

As in Sect. 2 of [6] (see also the appendix to [17]), we represent the dependence of the solution v to the problem (4.44), (4.45) on the forcing function u by an integral operator $\mathscr{I}$, writing $v = \mathscr{I}u$. We have the integral identities

$$(\mathscr{I}u, Lu) = (v, Lu) = (v, LHv). \tag{4.54}$$

A good choice of the coefficients a, m, and μ in the operator H on the right-hand side of this identity will allow us to derive an energy inequality, which will be used to prove weak existence via Theorem 2.2 (the Riesz Representation Theorem).

The main existence result rests on an improvement in the fundamental inequality for distribution solutions. It is convenient to begin by establishing a second multiplier identity (c.f. Sect. 2.4.3).

Proposition 4.2 ([24]). *Define the operator L on functions $v \in C^3(\Omega)$ with $v \equiv 0$ on $\partial\Omega$, by*

$$Lv = \mathcal{K}(x, y)\, v_{xx} + v_{yy} + \kappa_1 v_x + \kappa_2 v,$$

where the type-change function $\mathcal{K}$ and the lower-order coefficients $\kappa_1 = \kappa_1(x, y)$, $\kappa_2 = \kappa_2(x, y)$ are all C^3 functions. Define

$$Hv = av + bv_x + cv_y,$$

where a is a constant; c is a linear function of x and y; b is linear in x but possibly nonlinear in y; b and c have vanishing mixed partial derivatives and are assumed to be sufficiently smooth in x and y. Then

$$\int\int_\Omega v \cdot LHv\, dx dy = \frac{1}{2} \oint_{\partial\Omega} \left(\mathcal{K} v_x^2 + v_y^2\right)(c\, dx - b\, dy)$$

$$+ \int\int_\Omega \omega v^2 + \alpha v_x^2 + 2\beta v_x v_y + \gamma v_y^2 dx dy,$$

where

$$2\omega = (\mathcal{K}_{xx} - \kappa_{1x} + 2\kappa_2)\, a - [(\mathcal{K}_{xx} - \kappa_{1x} + \kappa_2)\, b]_x$$

$$- [(\mathcal{K}_{xx} - \kappa_{1x} + \kappa_2)\, c]_y\, ;$$

$$\alpha = \left(\frac{c_y - b_x}{2} - a\right)\mathcal{K} + \left(\frac{3}{2}\mathcal{K}_x - \kappa_1\right) b + \frac{c}{2}\mathcal{K}_y\, ;$$

$$2\beta = (\mathcal{K}_x - \kappa_1)\, c - (c_x \mathcal{K} + b_y)\, ;$$

$$\gamma = \frac{b_x - c_y}{2} - a.$$

Proof. Writing

$$LHv = \mathcal{K}\left[(a + 2b_x)\, v_{xx} + b v_{xxx} + 2c_x v_{yx} + c v_{yxx}\right]$$

$$+ (a + 2c_y)\, v_{yy} + b_{yy} v_x + 2b_y v_{xy} + b v_{xyy} + c v_{yyy}$$

$$+ \kappa_1\left[(a + b_x)\, v_x + b v_{xx} + c_x v_y + c v_{yx}\right] + \kappa_2\left(av + bv_x + cv_y\right),$$

we have

$$v \cdot LHv = \sum_{i=1}^{16} \tau_i,$$

where

$$\tau_1 = v\mathcal{K}\,(a + 2b_x)\,v_{xx} = \left\{\left[(a + 2b_x)\left(\mathcal{K}v_x - \frac{1}{2}\mathcal{K}_x v\right)\right]v\right\}_x$$

$$-\mathcal{K}\,(a + 2b_x)\,v_x^2 + \frac{1}{2}\mathcal{K}_{xx}\,(a + 2b_x)\,v^2;$$

$$\tau_2 = v\mathcal{K}bv_{xxx}$$

$$= \left\{-\frac{1}{2}\mathcal{K}bv_x^2 + \left[b\left(\mathcal{K}v_{xx} - \mathcal{K}_x v_x + \frac{1}{2}\mathcal{K}_{xx}v\right) + b_x\left(\mathcal{K}_x v - \mathcal{K}v_x\right)\right]v\right\}_x$$

$$-\frac{1}{2}\left(\mathcal{K}_{xxx}b + 3\mathcal{K}_{xx}b_x\right)v^2 + \frac{3}{2}\left(\mathcal{K}b\right)_x v_x^2;$$

$$\tau_3 = 2v\mathcal{K}c_x v_{yx} = 2\left(v\mathcal{K}c_x v_y\right)_x - \left(\mathcal{K}_x c_x v^2\right)_y - 2\mathcal{K}c_x v_x v_y + \mathcal{K}_{xy}c_x v^2;$$

$$\tau_4 = v\mathcal{K}cv_{yxx}$$

$$= \left\{v\left[c\left(\mathcal{K}v_{yx} - \mathcal{K}_x v_y\right) - \mathcal{K}c_x v_y\right]\right\}_x - \left\{\frac{1}{2}\mathcal{K}cv_x^2 - \left(\frac{1}{2}\mathcal{K}_{xx}c + \mathcal{K}_x c_x\right)v^2\right\}_y$$

$$-\frac{1}{2}\left\{\left[\mathcal{K}_{xxy}c + \mathcal{K}_{xx}c_y + 2\mathcal{K}_{xy}c_x\right]v^2 - \left(\mathcal{K}c\right)_y v_x^2\right\} + \left(\mathcal{K}c\right)_x v_x v_y;$$

$$\tau_5 = v\left(a + 2c_y\right)v_{yy} = \left[v\left(a + 2c_y\right)v_y\right]_y - \left(a + 2c_y\right)v_y^2;$$

$$\tau_6 = vb_{yy}v_x = \frac{1}{2}\left(b_{yy}v^2\right)_x;$$

$$\tau_7 = 2vb_y v_{xy} = \left(2vb_y v_x\right)_y - 2b_y v_x v_y - \left(b_{yy}v^2\right)_x;$$

$$\tau_8 = vbv_{xyy} = -\left(\frac{1}{2}bv_y^2\right)_x + \left[\left(bv_{xy} - b_y v_x\right)v\right]_y + \frac{1}{2}\left[b_x v_y^2 + \left(b_{yy}v^2\right)_x\right] + b_y v_x v_y;$$

$$\tau_9 = vcv_{yyy} = -\frac{1}{2}\left(cv_y^2\right)_y + \left[\left(cv_{yy} - c_y v_y\right)v\right]_y + \frac{3}{2}c_y v_y^2;$$

$$\tau_{10} = v\kappa_1\,(a + b_x)\,v_x = \frac{1}{2}\left[(a + b_x)\kappa_1 v^2\right]_x - \frac{1}{2}\kappa_{1x}\,(a + b_x)\,v^2;$$

$$\tau_{11} = v\kappa_1 bv_{xx} = \left\{\left[\kappa_1 bv_x - \frac{1}{2}\left(\kappa_1 b\right)_x v\right]v\right\}_x + \frac{1}{2}\left(\kappa_{1xx}b + 2\kappa_{1x}b_x\right)v^2 - \kappa_1 bv_x^2;$$

$$\tau_{12} = v\kappa_1 c_x v_y = \frac{1}{2}\left[\left(\kappa_1 c_x v^2\right)_y - \kappa_{1y}c_x\right]v^2;$$

$$\tau_{13} = v\kappa_1 cv_{yx} = \left(v\kappa_1 cv_y\right)_x - \frac{1}{2}\left[\left(\kappa_1 c\right)_x v^2\right]_y - \kappa_1 cv_x v_y + \frac{1}{2}\left[\kappa_{1y}c_x + \left(\kappa_{1x}c\right)_y\right]v^2;$$

$$\tau_{14} = \kappa_2 av^2;$$

$$\tau_{15} = v\kappa_2 b v_x = \frac{1}{2}\left(\kappa_2 b v^2\right)_x - \frac{1}{2}\left(\kappa_2 b\right)_x v^2;$$

$$\tau_{16} = v\kappa_2 c v_y = \frac{1}{2}\left(c\kappa_2 v^2\right)_y - \frac{1}{2}\left(c\kappa_2\right)_y v^2.$$

Collect terms and integrate over Ω. Applying the Divergence Theorem, taking into account that v (but not necessarily v_x or v_y) vanishes on $\partial\Omega$, completes the proof. $\square$

Note that if (b, c) is a vector field on Ω, then the term $c\,dx - b\,dy$ in the boundary integral of Proposition 4.2 can be written

$$c\,dx - b\,dy = (b, c) \cdot \hat{\mathbf{n}},$$

where $\hat{\mathbf{n}}$ is the unit normal to $\partial\Omega$, oriented so that if the boundary is traversed in a positive direction, the interior of Ω is on the left; c.f. (3.18) of [15] or (2.4) of [18].

Proposition 4.2 is different in its details from the integration by parts formula, Proposition 2.3, which was used in the proof of Lemma 3.1. For example, the coefficients of the operator H are allowed to be multivariate (as is $\mathscr{K}$), but are constrained to be linear in certain of their arguments, and the operator L has a more general distribution of lower-order terms. This last difference means that in general the terms in the second derivative of $\mathscr{K}$ do not cancel, as they did in the proof of Lemma 3.1. Although we apply Proposition 4.2 to a formally self-adjoint operator in the following lemma, we will use a non-self-adjoint form of the proposition in Lemma 4.2 of Sect. 4.5.

Now we prove the fundamental inequality:

Lemma 4.1. *Suppose that x is non-negative on Ω and that the origin of coordinates lies on $\partial\Omega$. Let Ω be star-shaped with respect to the flow of the vector field $V = -(b, c)$ for $b = mx$ and $c = \mu y$, where m and μ are positive constants and m exceeds 3μ. Then there exists a positive constant C for which the inequality*

$$||u||_{L^2(\Omega;|\mathscr{K}|)} \le C||Lu||_{H^{-1}(\Omega;\mathscr{K})} \tag{4.55}$$

holds for every $u \in C_0^\infty(\Omega)$, where $\mathscr{K}(x, y) = x - y^2$ and L is defined by (4.42).

Proof. Let v satisfy (4.44), (4.45) on Ω for $a = -M$, where M is a positive number satisfying

$$M = \frac{m - 3\mu}{2} - \delta$$

for some sufficiently small positive number δ. Integrate the integral identities (4.54) by parts, using Proposition 4.2 and the compact support of u. We have

$$\int\int_\Omega v \cdot LHv\,dx\,dy = \frac{1}{2}\oint_{\partial\Omega}\left(\mathscr{K}v_x^2 + v_y^2\right)(c\,dx - b\,dy)$$

$$+ \int\int_\Omega \alpha v_x^2 + \gamma v_y^2\,dx\,dy, \tag{4.56}$$

where

$$\alpha = \mathcal{K}\left(\frac{c_y - b_x}{2} - a\right) + \frac{1}{2}b + \frac{1}{2}\mathcal{K}_y c$$

$$= \left(\frac{m}{2} - \mu - \delta\right)x + \delta y^2$$

and

$$\gamma = -a - \frac{c_y}{2} + \frac{b_x}{2} = M - \frac{\mu - m}{2} = m - 2\mu - \delta > \mu - \delta.$$

On the *elliptic* region Ω^+, $K > 0$ and

$$\left(\frac{m}{2} - \mu - \delta\right)x > \left(\frac{\mu}{2} - \delta\right)x \geq \delta x$$

provided we choose δ so small that $\mu/4 \geq \delta$. Then on Ω^+,

$$\alpha \geq \delta\left(x + y^2\right) \geq \delta\left(x - y^2\right) = \delta\mathcal{K} = \delta|\mathcal{K}|.$$

On the *hyperbolic* region Ω^-, $\mathcal{K} < 0$ and

$$\alpha = \left(\frac{m}{2} - \mu\right)x + \delta\left(y^2 - x\right) \geq \frac{\mu}{2}x + \delta\left(-\mathcal{K}\right) \geq \delta|\mathcal{K}|.$$

We show that the integrand of the boundary integral in (4.56) is non-negative: As we explained in the Remark following the proof of Proposition 4.1, the form of (4.44), (4.45) implies that v vanishes identically in a sufficiently small neighborhood of each point of $\partial\Omega \setminus \{(0,0)\}$. This implies that $v_x^2 = v_y^2 = 0$ in a sufficiently small neighborhood of $\partial\Omega \setminus \{K = 0\}$. That in turn implies that

$$Kv_x^2 + v_y^2 = 0$$

in a sufficiently small neighborhood of $\partial\Omega \setminus \{K = 0\}$. On the curve $K = 0$, we have $x = y^2$ and $dx = 2y\,dy$, implying that

$$cdx - bdy = \left(2\mu y^2 - mx\right)dy = (2\mu - m)x\,dy.$$

But $m > 3\mu > 2\mu$, $x \geq 0$ on $\overline{\Omega}$, and $dy \leq 0$ on K; so on the resonance curve,

$$v_y^2\left(cdx - bdy\right) \geq 0.$$

Thus the integrand of the boundary integral in (4.56) is bounded below by zero. We find that if δ is sufficiently small relative to μ, then

$$(v, LHv) \geq \delta \int\!\!\int_\Omega \left(|\mathcal{K}|v_x^2 + v_y^2\right)dx\,dy. \tag{4.57}$$

The upper estimate is immediate, as

$$(v, LHv) = (v, Lu) \leq \|v\|_{H_0^1(\Omega;\mathcal{K})} \|Lu\|_{H^{-1}(\Omega;\mathcal{K})}. \tag{4.58}$$

Combining (4.57) and (4.58), we obtain

$$\|v\|_{H_0^1(\Omega;\mathcal{K})} \leq C \|Lu\|_{H^{-1}(\Omega;\mathcal{K})}. \tag{4.59}$$

The assertion of Lemma 4.1 now follows from (4.44) by the continuity of H as a map from $H_0^1(\Omega;\mathcal{K})$ into $L^2(\Omega;|\mathcal{K}|)$. This completes the proof of Lemma 4.1. $\square$

Theorem 4.2 ([26]). *Let Ω be star-shaped with respect to the flow of the vector field $-V = (mx, \mu y)$, where m and μ are defined as in Lemma 4.1. Suppose that x is nonnegative on Ω and that the origin of coordinates lies on $\partial\Omega$. Then for every $f \in L^2(\Omega;|\mathcal{K}|^{-1})$ there is a unique weak solution $u \in H_0^1(\Omega;\mathcal{K})$ to the Dirichlet problem (4.42), (4.43) where $\mathcal{K} = x - y^2$.*

Proof. The proof follows the outline of the arguments in [15], Sect. 3. Defining a linear functional J_f by the formula

$$J_f(L\xi) = (f, \xi), \quad \xi \in C_0^\infty(\Omega),$$

we estimate

$$|J_f(L\xi)| \leq \|f\|_{L^2(\Omega;|\mathcal{K}|^{-1})} \|\xi\|_{L^2(\Omega;|\mathcal{K}|)} \leq C\|f\|_{L^2(\Omega;|\mathcal{K}|^{-1})} \|L\xi\|_{H^{-1}(\Omega;\mathcal{K})}, \tag{4.60}$$

by applying Lemma 4.1 to ξ. Thus J_f is a bounded linear functional on the subspace of $H^{-1}(\Omega;\mathcal{K})$ consisting of elements having the form $L\xi$ with $\xi \in C_0^\infty(\Omega)$. As we have done previously, we extend J_f to the closure of this subspace by Theorem 2.1 and use Theorem 2.2 to obtain the existence of an element $u \in H_0^1(\Omega;\mathcal{K})$ for which

$$\langle u, L\xi \rangle = (f, \xi),$$

where $\xi \in H_0^1(\Omega;\mathcal{K})$. There exists a unique, continuous, self-adjoint extension $L : H_0^1(\Omega;\mathcal{K}) \to H^{-1}(\Omega;\mathcal{K})$. As a result, if a sequence u_n of smooth, compactly supported functions approximates u in the norm $H_0^1(\Omega;\mathcal{K})$, then Lu_n converges in norm to an element $\tilde{f}$ of $H^{-1}(\Omega;\mathcal{K})$. Taking the limit

$$\lim_{n\to\infty} \langle u - u_n, L\xi \rangle = \left(f - \tilde{f}, \xi \right),$$

we conclude that, because the left-hand side vanishes for all $\xi \in H_0^1(\Omega;\mathcal{K})$, the right-hand side must vanish as well. This proves the existence of a weak solution. Taking the difference of two weak solutions, we find that this difference is zero in $H_0^1(\Omega;\mathcal{K})$ by Lemma 4.1, the linearity of L, and Proposition 2.2 (c.f. [15]). This completes the proof of Theorem 4.2. $\square$

It was observed in [21] that mathematical reasoning and physical reasoning were at odds regarding the cold plasma model: physical reasoning suggested that the closed Dirichlet problem should be well posed for the model equation (4.42), whereas the analysis in [21] suggests that the closed Dirichlet problem is over-determined for at least one version of the model equation on a typical domain. This apparent contradiction is resolved for the case of a formally self-adjoint model in Theorem 4.2.

Moreover, the choice of weight class supports the physical reasoning. The weight function in Theorem 4.2 vanishes on the resonance curve $y = x^2$. In Corollary 6 of [24] an existence theorem for the boundary value problem of Theorem 4.2, without uniqueness, was demonstrated in a weighted function space in which the weight function vanished on the line $y = 0$. Thus in each case the weight function vanishes at the origin of coordinates, which in the physical model is a point of tangency between the resonance curve and the flux line $x = 0$. These results support the conjecture in [30] of a plasma heating zone at that point (c.f. Fig. 4.1), which would take the mathematical form of a singularity in solutions to the equations.

The conjecture of a singular point at the origin is also supported by an analysis of characteristic lines. We ask what must happen algebraically in order for a characteristic line to pass through the origin. The values of x and y would need to satisfy the equation

$$x = \lambda y^2 \tag{4.61}$$

for some constant λ, as well as the characteristic equation

$$\left(x - y^2\right) dy^2 + dx^2 = 0 \tag{4.62}$$

for (4.35). Substituting (4.61) into (4.62) yields the equation

$$\frac{dy^2}{(2\lambda y dy)^2} = \frac{1}{(1 - \lambda) y^2}, \tag{4.63}$$

which reduces to the quadratic equation

$$4\lambda^2 + \lambda - 1 = 0.$$

This polynomial has two real solutions. Both solutions are exceeded by one, so $x < y$ whenever (4.61) is satisfied. Because the characteristic equation for (4.35) has two roots at the origin, four characteristic lines must pass through the origin. That is two more characteristic lines passing through the origin than through any other hyperbolic point. So we expect solutions of equations having the form (4.35) to be singular at the origin [21].

The mathematical restriction in Theorem 4.2 that the points of Ω may not lie in the negative half-plane corresponds a physical requirement that boundary conditions be placed on one side of the flux line. As the resonance frequency is naturally restricted to the same side of that line, the mathematical requirement seems compatible with the physical model.

Fig. 4.6 A domain bounded
by a curve in the elliptic
region and two intersecting
characteristics in the first
quadrant

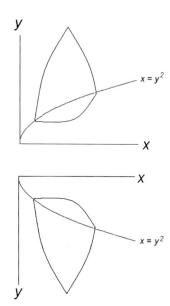

Fig. 4.7 A domain bounded
by a curve in the elliptic
region and two intersecting
characteristics in the fourth
quadrant

In addition to its intrinsic mathematical and physical interest, the formulation of boundary value problems illuminates other topics in the analysis of the cold plasma model. For example, it is implied by the tedious analytic argument of Theorem 4.1 that, away from the origin, the governing equation for the model is of Tricomi type, whereas in any neighborhood of the origin it is of Keldysh type. This distinction is also suggested, without reference to such terminology, by other analytic arguments in [27] and in Sect. 4 of [31]. If we try to form a standard elliptic–hyperbolic boundary value problem in which the hyperbolic boundary is composed of intersecting characteristics, we might choose both these characteristics to originate at points on the arc of the resonance curve $x = y^2$ that lies in the first quadrant, or both of them to lie in the fourth quadrant. We then obtain a standard problem for a vertical-ice-cream-cone-shaped region (Figs. 4.6 and 4.7).

The resulting boundary value problems are similar to those formulated for the Tricomi equation ((3.18) with $\mathscr{K}(\xi, \eta) = \xi$). The domain geometry in Figs. 4.6 and 4.7 is exactly analogous to Fig. 3.1 of Sect. 3.1; see also Fig. 2 of [20], with the line AB in that figure replaced by an arc of the curve $x = y^2$, lying either completely above or completely below the x-axis. But the origin will not be included, as that is a singular point of the characteristic equation (4.62). If we include the origin, we are led to a hyperbolic region bounded by characteristics that lie partly in the second and third quadrants (Fig. 4.8).

This horizontal-ice-cream-cone-shaped region is similar to those formulated for Cinquini-Cibrario's equation (3.4). In that case the typical domain geometry is analogous to Fig. 3.2 of Sect. 3.3 or Fig. 3.4 of Sect. 3.5.2. A similar domain is shown in Fig. 2 of [4], if the line MN in that figure is replaced by an arc of the

Fig. 4.8 A domain bounded
by a curve in the elliptic
region and two intersecting
characteristics, one in the first
and second quadrants and the
other in the third and fourth
quadrants

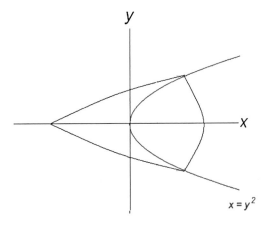

curve $x = y^2$ which is symmetric about the x-axis; see also Remark *i)* following
Corollary 11 of [24]. Thus the defining analytic character of the equation is clearly
apparent in the geometry of the natural boundary value problems.

Remarks. Recall that (4.42) was derived under the assumption that the zeroth-
order terms of (4.12) can be neglected. As Bers observes [3], there is a significant
mathematical difference between assuming that certain terms in a solution of
a partial differential equation are small, and assuming that certain terms in the
equation itself are small. Although in this case the mathematical hypothesis is
derived from physical intuition, it is reasonable to ask what effect, if any, the
assumption has on the truth value of Theorem 4.2. The missing zeroth-order terms
of (4.12) are (the real part of)

$$ik_2 \left(K_{12x} + K_{22y} + K_{32z} \right) \varphi e^{ik_2 y} = ik_2 \left(s_y - i d_x \right) ik_2 \varphi e^{ik_2 y} \equiv \kappa_0 \varphi e^{ik_2 y}.$$

We assume that κ_0 and its first derivatives are bounded above almost everywhere by
a sufficiently small positive number δ'. In that case, we can show that Theorem 4.2
remains valid in the presence of such zeroth-order terms by replacing the operator
L acting on ξ by the perturbed operator

$$\tilde{L}\xi = (L + \kappa_0) \xi$$

and proceeding as in the proof of Theorem 4.2, using Proposition 4.2 with $\kappa_1 = 1$
and $\kappa_2 = \kappa_0$. We find that

$$\left(v, \tilde{L} H v \right) \geq \delta \int\!\!\int_\Omega \left(\mathcal{H} v_x^2 + v_y^2 \right) dx dy - \int\!\!\int_\Omega \left[2\kappa_0 a - (\kappa_0 b)_x - (\kappa_0 c)_y \right] v^2 dx dy,$$

where

$$\left| 2\kappa_0 a - (\kappa_0 b)_x - (\kappa_0 c)_y \right| \leq \delta' \left[2|a| + (|m| + |\mu|) \left(1 + |\Omega| \right) \right] \equiv \mathscr{C} \delta'.$$

This implies the inequality

$$\left(v, \tilde{L}Hv\right) \geq \delta \int \int_{\Omega} \left(\mathcal{K}v_x^2 + v_y^2\right) dxdy - \mathcal{C}\delta' \int \int_{\Omega} v^2 dxdy$$

$$\geq \delta \int \int_{\Omega} \left(\mathcal{K}v_x^2 + v_y^2\right) dxdy - \mathcal{C}' \int \int_{\Omega} v_y^2 dxdy,$$

where the constant

$$\mathcal{C}' \equiv \mathcal{C}\delta' |\Omega|^2$$

arises from standard path integration arguments analogous, for example, to the proof of Proposition 2.2. We conclude that

$$\left(v, \tilde{L}Hv\right) \geq \left(\delta - \mathcal{C}'\right) \int \int_{\Omega} \left(\mathcal{K}v_x^2 + v_y^2\right) dxdy, \tag{4.64}$$

where $\delta - \mathcal{C}'$ exceeds zero provided δ' is sufficiently small with respect to the other parameters. But also,

$$\left(v, \tilde{L}Hv\right) = \left(v, \tilde{L}u\right) \leq \|v\|_{H_0^1(\Omega;\mathcal{K})} \|\tilde{L}u\|_{H^{-1}(\Omega;\mathcal{K})}. \tag{4.65}$$

Combining (4.64) and (4.65), we obtain

$$\|v\|_{H_0^1(\Omega;\mathcal{K})} \leq C \|\tilde{L}u\|_{H^{-1}(\Omega;\mathcal{K})}.$$

Apply (4.44), using the continuity of H as a map from $H_0^1(\Omega;\mathcal{K})$ into $L^2(\Omega;|\mathcal{K}|)$. We obtain the fundamental inequality (4.55).

This estimate does not settle the issue, as the terms in (4.12) are subjected to certain rescalings and choices of parameter in [27] in the course of deriving an equation qualitatively similar to (4.42). And in fact, (4.42) is only a toy model for the equation governing fully electromagnetic waves in zero-temperature plasma, in which many terms have been neglected or simplified (Sect. 4.1.2). Given the delicate dependence of existence theorems for Keldysh-type equations on the form of the lower-order terms (see, e.g., [10]), the question of which terms in the cold plasma model can be neglected and which cannot remains quite unsettled.

4.4   Similarity Solutions

We are not able to show that weak solutions to the closed Dirichlet problem for equations having the form (4.42) are in $H_{loc}^1(\Omega)$. There is numerical evidence that they are not [21]. However, for a special class of solutions, we can give a very rough estimation of the weight class necessary to achieve a finite-energy solution if the energy functional is a weighted $H^{1,2}$-norm.

Consider weighted energy functionals of the standard form similar to those of [7, 11] (rather than in the special form introduced in Sect. 2.4.2), that is

$$E = \int_{\Omega} |\mathscr{K}| \left(u_x^2 + u_y^2 \right) dxdy.$$

In a narrow vertical lens about the origin – which is the point of interest, $|x|$ will be small but $|y|$ may not be. In such a lens, the energy will resemble

$$E_2 \equiv \int_{\Omega} y^2 \left(u_x^2 + u_y^2 \right) dxdy.$$

More generally, we may consider weighted energies having the form

$$E_h = \int_{\Omega} |y|^h \left(u_x^2 + u_y^2 \right) dxdy. \tag{4.66}$$

Such energies were considered for the cold plasma model, without any smallness assumption on $|x|$, in [23] and in Sect. 4 of [24]. If $h > 0$ is fixed, we can derive conditions on a class of solutions sufficient to guarantee finite energy.

4.4.1 Reduction to an Ordinary Differential Equation

Consider the equation

$$\left(x - y^2 \right) u_{xx} + u_{yy} = 0 \tag{4.67}$$

under the coordinate transformation (3.94) – that is,

$$w = t^2 x, \quad z = ty, \quad t > 0.$$

Then $u_x = u_w w_x = t^2 u_w$ and $u_y = u_z z_y = t u_z$, from which we obtain

$$u_{xx} = t^2 u_{ww} w_x = t^4 u_{ww}$$

and

$$u_{yy} = t u_{zz} z_y = t^2 u_{zz}.$$

Because $x = t^{-2}w$ and $y = t^{-1}z$, we obtain

$$0 = \left(x - y^2 \right) u_{xx} + u_{yy} = t^2 \left[\left(w - z^2 \right) u_{ww} + u_{zz} \right].$$

Like Cinquini-Cibrario's equation, (4.67) is invariant under rescalings of the form (3.94). We could apply the solution template (3.95) that was applied in Sect. 3.7 to (3.4). However, in order to recover results in the previous literature on the cold plasma model, we look for similarity solutions having the form

$$\frac{y^2}{x} = \frac{t^{-2}z^2}{t^{-2}w} = \frac{z^2}{w}. \tag{4.68}$$

Following particularly [30], (82) (see also [23], (2.15)), we let

$$\varphi = x^\nu F\left(\frac{y^2}{x}\right). \tag{4.69}$$

Then

$$\varphi_x = \nu x^{\nu-1} F - y^2 x^{\nu-2} F';$$

$$\varphi_{xx} = x^{\nu-2}\left[\nu(\nu-1)F - 2\frac{y^2}{x}(\nu-1)F' + \frac{y^4}{x^2}F''\right];$$

$$\varphi_y = 2yx^{\nu-1}F';$$

$$\varphi_{yy} = 2x^{\nu-1}F' + 4y^2 x^{\nu-2}F''.$$

Equation (4.67) assumes the form $L\varphi = 0$ for

$$L\varphi = x^{\nu-2}\left\{(x-y^2)\left[\nu(\nu-1)F - 2\frac{y^2}{x}(\nu-1)F' + \frac{y^4}{x^2}F''\right] + 2xF' + 4y^2 F''\right\}.$$

Taking into account (4.68), let

$$\mu = \frac{y^2}{x}.$$

(Note that μ is undefined at the origin; but on the resonance curve, $\mu = 1$ and is nonsingular.) We obtain

$$x^{1-\nu}L\varphi = (1-\mu)\left[\nu(\nu-1)F(\mu) - 2\mu(\nu-1)F'(\mu) + \mu^2 F''(\mu)\right]$$
$$+2F'(\mu) + 4\mu F''(\mu) = 0; \tag{4.70}$$

c.f. [30], (83). Equation (4.70) is sufficiently regular to be transformable into a standard hypergeometric equation – c.f. Theorem 8.1 of [22], Chap. 5 – so there is no question about the existence of the function F.

4.4.2   Computing the Weight Class

Given that F exists, we can approximate it by μ^ν (following [30]). Dividing through by the factor $1 - \mu$, the left-hand side of (4.70) is approximately

$$\left[\nu(\nu-1)\mu^\nu - 2(\nu-1)\mu\nu\mu^{\nu-1} + \mu^2\nu(\nu-1)\mu^{\nu-2}\right]$$
$$+\frac{1}{1-\mu}\left[2\nu\mu^{\nu-1} + 4\mu\nu(\nu-1)\mu^{\nu-1}\right].$$

If μ sufficiently exceeds one (that is, if $|x|$ is sufficiently small relative to y^2), then we can neglect the second bracketed term to obtain

$$\mu^\nu \left[\nu (\nu - 1) - 2\nu (\nu - 1) + \nu (\nu - 1) \right] = 0.$$

So the choice $F(\mu) = \mu^\nu$ approximately satisfies (4.70) in the region of small $|x|$.
 Substituting this choice into (4.69), we obtain

$$u(x, y) \approx \varphi = x^\nu F(\mu) = x^\nu \mu^\nu = x^\nu \left(\frac{y^2}{x} \right)^\nu = y^{2\nu}.$$

As long as Ω is finite in the y direction, the weighted energy E_h given by (4.66) will be finite whenever we choose ν so that

$$E_h = \int_\Omega |y|^h \left(u_x^2 + u_y^2 \right) dxdy = \int_\Omega |y|^h \left[\partial_y \left(y^{2\nu} \right) \right]^2 dxdy < \infty,$$

or $\nu \geq (1 - h)/4$. In particular, if $\nu = (2 - h)/4$, then E_h is proportional to the volume of Ω.

4.5 Will These Methods Work for Cinquini-Cibrario's Equation?

Cinquini-Cibrario's equation arises in the cold plasma model as the case in which a flux line is collinear with a resonance curve. An electromagnetic potential leading to such a geometry could be induced by applying a driving potential to the metallic plates of a condenser (Fig. 4.9). Moreover, the approximate collinearity of the flux line and the resonance curve could arise under certain extreme conditions in a tokamak (for very high fields or very low densities – see p. 47 of [30]). But the relation to Cinquini-Cibrario's equation of such models is mainly formal, as in this kind of "slab" geometry the governing equation is typically reduced to an ordinary differential equation; see the Appendix to [14] or Sect. C of [9].

 There is, however, mathematical interest in applying the methods of this chapter to Cinquini-Cibrario's equation, and it is natural to wonder whether the delicate methods used to establish the existence of weak solutions in the cold plasma model also work for Cinquini-Cibrario's equation.

 Corresponding to our use of (4.45) in the preceding section, we solve the Cauchy problem for an integral multiplier having the form

$$Mv = av + bv_x + cv_y = u \text{ in } \Omega,$$

$$v = 0 \text{ in } \Gamma, \tag{4.71}$$

Fig. 4.9 An electrostatic
oscillation is produced
between metallic plates
surrounding the plasma.
Adapted from [14]

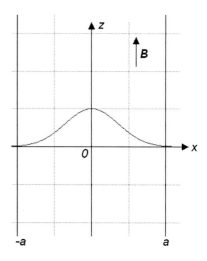

Fig. 4.9 An electrostatic
oscillation is produced
between metallic plates
surrounding the plasma.
Adapted from [14]

where $\Gamma \in \partial\Omega$ is the boundary arc on which data have been assigned. Choose
$b = mx$ and $c = \mu y$, where μ is a positive constant and m is a step function which
jumps at the line $x = 0$ and has the same sign as μ. Take a to be a constant of
sufficiently large magnitude having the sign opposite to that of μ and m.

As in [15], a *weak solution* to the Dirichlet problem (3.35), (3.36) is a function
$u \in H_0^1(\Omega; \mathcal{K})$ for which

$$\langle Lu, \xi \rangle = \langle f, \xi \rangle$$

for every $\xi \in H_0^1(\Omega; \mathcal{K})$, where L satisfies (3.7), (3.9) and (3.10). Integrating by
parts as in [15],

$$\langle Lu, \xi \rangle \equiv \int\int_{\Omega} \left([\mathcal{K}(x)u_x]_x + u_{yy} \right) \xi \, dxdy = -\int\int_{\Omega} \left(\mathcal{K} u_x \xi_x + u_y \xi_y \right) dxdy.$$

The right-hand side requires L to be formally self-adjoint, which fixes the first-order
term. As in the cold plasma model, the existence of a weak solution is equivalent to
the existence of a sequence $u_n \in C_0^{\infty}(\Omega)$ such that

$$||u_n - u||_{H_0^1(\Omega; \mathcal{K})} \to 0 \text{ and } ||Lu_n - f||_{H^{-1}(\Omega; \mathcal{K})} \to 0$$

as n tends to infinity.

Estimating the coefficients α, β, and γ as in the cold-plasma case, we find that
$\beta = 0$, and that

$$\alpha = \mathcal{K}(x) \left\{ -a + \frac{\mu}{2} + \left[\frac{x\mathcal{K}'(x)}{\mathcal{K}(x)} - 1 \right] \frac{m}{2} \right\}.$$

We have

$$\lim_{x \to 0} \frac{x \mathscr{K}'(x)}{\mathscr{K}(x)} = \lim_{x \to 0} \frac{x \mathscr{K}'(x)}{[\mathscr{K}(x) - \mathscr{K}(0)]} = \frac{\mathscr{K}'(x)}{\mathscr{K}'(x)} = 1, \qquad (4.72)$$

as $\mathscr{K}(0) = 0$. In order for (2.12) to change type at $x = 0$, $\mathscr{K}(x)$ must be monotonic in at least a small interval about $x = 0$. Initially, suppose that $\mathscr{K}'(x)$ is positive near $x = 0$. The sum $-a + \mu/2$ cannot be zero, as a and μ have been given opposite sign in order to obtain the existence of a solution to (4.71). If $-a + \mu/2$ is positive, then for small negative values of x, $\mathscr{K}(x)$ will be negative. The contribution of $[(x \mathscr{K}'/\mathscr{K}) - 1] m$ will be small by (4.72), so α will be negative. If $-a + \mu/2$ is negative, then α will be negative for small positive values of x for the same reason. Essentially the same argument holds in the case for which $\mathscr{K}'(x)$ is negative near $x = 0$ [24].

Thus the arguments for the unique existence of weak solutions to the closed Dirichlet problem do not appear to extend in an *obvious* way to Cinquini-Cibrario's equation – or, more generally, to formally self-adjoint equations satisfying (3.7), (3.9), and (3.10). (Of course they may extend in a way that is not obvious.)

However, we can apply the methods of Sect. 4.3 to a slightly generalized Cinquini-Cibrario equation in a straightforward way if we do not insist upon the uniqueness of the solution.

Lemma 4.2 ([24]). *Consider an equation having the form*

$$Lu \equiv x^{2k+1} u_{xx} + u_{yy} + c_1 x^{2k} u_x + c_2 u = 0, \qquad (4.73)$$

where $k \in \mathbf{Z}^+$ and the constants c_1 and c_2 satisfy $c_1 < k + 1$ and $c_2 < 0$ with $|c_2|$ sufficiently large. Let a portion of the line $x = 0$ lie in Ω and let the point $(0,0)$ lie on $\partial\Omega$. Assume that Ω is star-shaped with respect to the vector field $V = -(b, c)$, where $b = mx$, and $c = \mu y$. Let μ be a positive constant and let

$$m = \begin{cases} -a/\ell + \mu/2\ell - \delta/\ell \ \text{in} \ \Omega^+ \\ -a/\ell + \mu/2\ell + \delta/\ell \ \text{in} \ \Omega^- \end{cases}$$

for a positive constant δ, where $\ell = k + 1 - c_1$. Let a be a negative constant of sufficiently large magnitude. In particular, let a have sufficiently large magnitude that m is positive. Then for every $w \in C_0^\infty(\Omega)$ there exists a positive constant C for which

$$||w||_{L^2(\Omega; |K|)} \leq C ||L^* w||_{H^{-1}(\Omega; K)},$$

where $\mathscr{K} = x^{2k+1}$ and L^* is the formal adjoint of the differential operator L of (4.73) acting on w.

Proof. The proof is similar to that of Lemma 4.1. In Proposition 4.2, replace Lv by

$$L^* w = x^{2k+1} w_{xx} + w_{yy} + (4k + 2 - c_1) x^{2k} w_x$$
$$+ \left[2k (2k + 1 - c_1) x^{2k-1} + c_2 \right] w.$$

Initially we perform all operations over Ω^+ and Ω^- individually. On the interior of these sub-domains the coefficients are all smooth. We find that

$$\omega = M x^{2k-1} + c_2 (a - m/2 - \mu/2),$$

where M is a constant that depends on k, a, m, and c_1 but not on c_2. Because $a < 0$, $m > 0$, and $\mu > 0$, we conclude that ω is positive provided c_2 is a negative number having sufficiently large magnitude relative to the quantity $M/|a - m/2 - \mu/2|$. In addition, Proposition 4.2 implies that

$$
\begin{aligned}
\alpha &= -\mathcal{K}(a + 2b_x) + \frac{3}{2}(\mathcal{K}_x b + \mathcal{K} b_x) + \frac{1}{2}\mathcal{K} c_y - \kappa_1 b \\
&= x^{2k+1}\left[\frac{\mu}{2} - a - (k + 1 - c_1)m\right] = \delta|x|x^{2k};
\end{aligned}
\tag{4.74}
$$

$$
\beta = \frac{1}{2}(\mathcal{K}_x - \kappa_1)c = \frac{1}{2}(c_1 - 2k - 1)\mu y x^{2k};
$$

and

$$
\begin{aligned}
\gamma &= -a - \frac{c_y}{2} + \frac{b_x}{2} \\
&= \begin{cases} -(1 + 1/2\ell)a + (1/4\ell - 1/2)\mu - \delta/2\ell & \text{in } \Omega^+ \\ -(1 + 1/2\ell)a + (1/4\ell - 1/2)\mu + \delta/2\ell & \text{in } \Omega^- \end{cases}.
\end{aligned}
$$

Then

$$\gamma \geq \delta \tag{4.75}$$

by the hypotheses on a. Unlike the preceding examples, in this case the coefficient β of the cross terms is non-zero. We can write β in the form

$$2\beta = (c_1 - 2k - 1)\mu y x^{k+1/2} x^{k-1/2}.$$

Then $\forall\, \varepsilon > 0$ and any numbers ξ and η, we have by Young's inequality

$$2\beta\xi\eta \geq -|c_1 - 2k - 1|\mu|y|\left(\varepsilon|x|x^{2k}\xi^2 + \frac{1}{\varepsilon}|x|^{2k-1}\eta^2\right).$$

Choose ε so small and $|a|$ so large that both the inequality

$$|c_1 - 2k - 1|\mu \max_{\Omega}|y|\varepsilon < \delta$$

and the inequality

$$\frac{1}{\varepsilon}\max_{\Omega}|x|^{2k-1} < -\left(1 + \frac{1}{2\ell}\right)a + \left(\frac{1}{4\ell} - \frac{1}{2}\right)\mu - \frac{\delta}{2\ell}$$

hold on Ω. We can do this because $|\Omega|$ is bounded and $k \geq 1$. Then, using (4.74) and (4.75), we find that

$$\int\int_\Omega \left(\alpha\xi^2 + 2\beta\xi\eta + \gamma\eta^2\right) dxdy \geq C \int\int_\Omega \left(|\mathcal{K}|\xi^2 + \eta^2\right) dxdy. \qquad (4.76)$$

The function $b(x)$ fails to be differentiable on the boundary between Ω^+ and Ω^-. But the coefficients of the boundary terms involving b_x either vanish on the line $K = 0$ or cancel out, and the remaining boundary terms are smooth. Thus we can integrate over Ω^+ and Ω^-, using the support of v and the Divergence Theorem, to obtain

$$\left(w, L^* H w\right) \geq C \int\int_\Omega \left(|\mathcal{K}|w_x^2 + w_y^2\right) dxdy,$$

where we have taken $\xi = w_x$ and $\eta = w_y$ in (4.76). The remainder of the proof is the same as that of Lemma 4.1.   □

We obtain by the usual arguments

Theorem 4.3 ([24]). *Let Ω and V be defined as in Lemma 4.2. Then for every $f \in L^2\left(\Omega; |\mathcal{K}|^{-1}\right)$ there is a distribution solution u to the Dirichlet problem (4.73), (3.36) lying in $H_0^1\left(\Omega; \mathcal{K}\right)$ for $\mathcal{K} = x^{2k+1}$.*

Although the application of the term *distribution solution* to a solution which lies in weighted H^1 may strike some as over-cautious, this solution does not satisfy our formal definition of weak solution in any obvious way; c.f. Appendix A.1. There is currently no evidence that these solutions actually lie in $H^1(\Omega)$.

References

1. Allis, W.P: Waves in a plasma, Mass. Inst. Technol. Research Lab. Electronics Quart. Progr. Rep., Vol. **54**, No. 5, Cambridge (1959)
2. Astrom, E.O.: Waves in an ionized gas. Arkiv. Fysik **2**, 443 (1950)
3. Bers, L.: Mathematical Aspects of Subsonic and Transonic Gas Dynamics. Wiley, New York (1958)
4. Cibrario, M.: Intorno ad una equazione lineare alle derivate parziali del secondo ordine di tipe misto iperbolico-ellittica. Ann. Sc. Norm. Sup. Pisa, Cl. Sci., Ser. 2 **3**(3, 4), 255–285 (1934)
5. Czechowski, A., Grzedzielski, S.: A cold plasma layer at the heliopause. Adv. Space Res. **16**, 321–325 (1995)
6. Didenko, V.P.: On the generalized solvability of the Tricomi problem. Ukrain. Math. J. **25**, 10–18 (1973)
7. Fabes, E., Kenig, C., Serapioni, R.: The local regularity of solutions of degenerate elliptic equations. Commun. Partial Diff. Equations **7**, 77–116 (1982)
8. Gilbarg, D., Trudinger, N.S.: Elliptic Partial Differential Equations of Second Order. Springer, Berlin (1983)
9. Grossman, W., Weitzner, H.: A reformulation of lower-hybrid wave propagation and absorption. Phys. Fluids, **27**, 1699–1703 (1984)

10. Gu, C.: On partial differential equations of mixed type in n independent variables. Commun. Pure Appl. Math. **34**, 333–345 (1981)
11. Heinonen, J., Kilpeläinen, T., Martio, O.: Nonlinear Potential Theory and Degenerate Elliptic Equations. Oxford University Press, Oxford (1993)
12. Killian, T.C.,Pattard, T., Pohl, T., Rost, J.M.: Ultracold neutral plasmas. Phys. Rep. **449**, 77–130 (2007)
13. W. S. Kurth, Waves in space plasmas, e–note (n.d.). http://www-pw.physics.uiowa.edu/plasma-wave/tutorial/waves.html. Cited 2 Aug 2011
14. Lazzaro, E., Maroli, C.: Lower hybrid resonance in an inhomogeneous cold and collisionless plasma slab. Nuovo Cim. **16B**, 44–54 (1973)
15. Lupo, D., Morawetz, C.S., Payne, K.R.: On closed boundary value problems for equations of mixed elliptic-hyperbolic type. Commun. Pure Appl. Math. **60**, 1319–1348 (2007)
16. Lupo, D., Morawetz, C.S., Payne, K.R.: Erratum: "On closed boundary value problems for equations of mixed elliptic-hyperbolic type," [Commun. Pure Appl. Math. **60**, 1319–1348 (2007)]. Commun. Pure Appl. Math. **61**, 594 (2008)
17. Lupo, D., Payne, K.R.: A dual variational approach to a class of nonlocal semilinear Tricomi problems. Nonlinear Differential Equations Appl. **6**, 247–266 (1999)
18. Lupo, D., Payne, K.R.: Critical exponents for semilinear equations of mixed elliptic-hyperbolic and degenerate types. Commun. Pure Appl. Math. **56**, 403–424 (2003)
19. McDonald, K.T.: An electrostatic wave. arXiv:physics/0312025v1
20. Morawetz, C.S.: Mixed equations and transonic flow. Rend. Mat. **25**, 1–28 (1966)
21. Morawetz, C.S., Stevens, D.C., Weitzner, H.: A numerical experiment on a second-order partial differential equation of mixed type. Commun. Pure Appl. Math. **44**, 1091–1106 (1991)
22. Olver, F.W.J.: Asymptotics and Special Functions. A K Peters, Natick (1997)
23. Otway, T.H.: A boundary-value problem for cold plasma dynamics. J. Appl. Math. **3**, 17–33 (2003)
24. Otway, T.H.: Energy inequalities for a model of wave propagation in cold plasma. Publ. Mat. **52**, 195–234 (2008)
25. Otway, T.H.: Mathematical aspects of the cold plasma model. In: Duan, J., Fu, X., Yang, Y. (eds.) Perspectives in Mathematical Sciences, pp. 181–210. World Scientific Press, Singapore (2010)
26. Otway, T.H.: Unique solutions to boundary value problems in the cold plasma model. SIAM J. Math. Anal. **42**, 3045–3053 (2010)
27. Piliya, A.D., Fedorov, V.I.: Singularities of the field of an electromagnetic wave in a cold anisotropic plasma with two-dimensional inhomogeneity. Sov. Phys. JETP **33**, 210–215 (1971)
28. Sitenko, A.G., Stepanov, K.N.: On the oscillations of an electron plasma in a magnetic field. Z. Eksp. Teoret. Fiz. [in Russian] **31**, 642 (1956) [Sov. Phys. JETP, **4**, 512 (1957)]
29. Tonks, L., Langmuir, I.: Oscillations of ionized gases. Phys. Rev. **33**, 195–210 (1929)
30. Weitzner, H.: "Wave propagation in a plasma based on the cold plasma model." Courant Inst. Math. Sci. Magneto-Fluid Dynamics Div. Report MF–103, August, 1984
31. Weitzner, H.: Lower hybrid waves in the cold plasma model. Commun. Pure Appl. Math. **38**, 919–932 (1985)
32. Yamamoto, Y.: Existence and uniqueness of a generalized solution for a system of equations of mixed type. Ph.D. thesis, Polytechnic University of New York (1994)

Chapter 5
Light Near a Caustic

In the cold plasma model the sonic curve is a parabola. In the physical model presented in this chapter the sonic curve is a circle, and the elliptic region of the governing equation surrounds the hyperbolic region. Thus we can prescribe Dirichlet data on a suitable closed curve lying entirely in the elliptic region and obtain an elliptic–hyperbolic boundary value problem. Eventually, we will construct such a problem and show that it possesses a weak solution. In the next chapter the sonic curve will also be a circle; but in that case the hyperbolic region of the governing equation will enclose the elliptic region, leading to a significant reduction in regularity for elliptic–hyperbolic Dirichlet problems.

The fundamental equations for the propagation of light in the classical model are Maxwell's equations. As in Sect. 3.6 and Chap. 4, we will study elliptic–hyperbolic equations that result from applying Maxwell's equations in an unusual context – in this case, in the presence of a smooth convex caustic. In the absence of electric charges and currents, Maxwell's equations for the electric field $\mathbf{E}$ and the magnetic field $\mathbf{B}$ have the differential (vector) form

$$\nabla \times \mathbf{E} = -\frac{\partial \mathbf{B}}{\partial t}, \ t \in \mathbf{R}^{+} \cup \{0\} \tag{5.1}$$

(*Faraday's Law of induction*);

$$\nabla \times \mathbf{B} = \frac{1}{c^2} \frac{\partial \mathbf{E}}{\partial t} \tag{5.2}$$

(*Ampere's Law*);

$$\nabla \cdot \mathbf{E} = 0 \tag{5.3}$$

(*Gauss' Law for Electricity*);

$$\nabla \cdot \mathbf{B} = 0 \tag{5.4}$$

(*Gauss' Law for Magnetism*). In (5.2), c is the speed of light in the medium; that is,

$$c = \frac{c_0}{\nu},$$

T.H. Otway, *The Dirichlet Problem for Elliptic-Hyperbolic Equations of Keldysh Type*, Lecture Notes in Mathematics 2043, DOI 10.1007/978-3-642-24415-5_5, © Springer-Verlag Berlin Heidelberg 2012

where c_0 is the speed of light in a vacuum and v is the refractive index of the medium.

Using the standard properties of the curl operator, we compute

$$\nabla \times \nabla \times \mathbf{E} = \nabla (\nabla \cdot \mathbf{E}) - \nabla^2 \mathbf{E}. \tag{5.5}$$

Applying (5.3) to the right-hand side of (5.5), we obtain

$$\nabla \times \nabla \times \mathbf{E} = -\nabla^2 \mathbf{E}. \tag{5.6}$$

Equation (5.1) implies that

$$\nabla \times \nabla \times \mathbf{E} = \nabla \times \left(-\frac{\partial \mathbf{B}}{\partial t}\right) = -\frac{\partial}{\partial t} (\nabla \times \mathbf{B}). \tag{5.7}$$

Applying (5.2) to the right-hand-side of (5.7) yields

$$\nabla \times \nabla \times \mathbf{E} = -\frac{1}{c^2} \frac{\partial^2 \mathbf{E}}{\partial t^2}. \tag{5.8}$$

Now substituting the right-hand side of (5.6) into the left-hand side of (5.8), we obtain

$$\nabla^2 \mathbf{E} = \frac{1}{c^2} \frac{\partial^2 \mathbf{E}}{\partial t^2}, \tag{5.9}$$

which is the vector wave equation. (A similar argument works for the magnetic field $\mathbf{B}$.) The components of $\mathbf{E}$ satisfy the scalar wave equation

$$c^{-2} v_{tt} = \Delta v \equiv \sum_{n=1}^{3} v_{x^n x^n}. \tag{5.10}$$

Here $(x^1, x^2, x^3) \equiv x \in \mathbf{R}^3$. In the simplest case, solutions will be stationary waves. These can be written in the form

$$v(x, t) = u(x)e^{i\omega t}, \tag{5.11}$$

where the constant ω is the angular frequency of the wave and $i^2 = -1$. Expressing $\exp[i\omega t]$ in terms of sines and cosines via Euler's formula, the function $u(x)$ acquires a physical interpretation as the amplitude of a wave vibrating in simple harmonic motion. Substituting (5.11) into (5.10) and using the chain rule, we find that

$$c^{-2} v_{tt} = -\left(\frac{\omega v}{c_0}\right)^2 e^{i\omega t} u = \sum_{n=1}^{3} v_{x^n x^n} = e^{i\omega t} \sum_{n=1}^{3} u_{x^n x^n}.$$

Selecting the terms in u from this sequence of identities, we obtain the *Helmholtz* (or *reduced wave*) equation

$$\Delta u + \kappa^2 u = 0, \tag{5.12}$$

where $\kappa = \omega v / c_0$.

As the *wave number* $k = \omega / c_0$ tends to infinity, we obtain the geometrical optics approximation, in which the equations themselves can be replaced by the Euclidean geometry of rays. In this range the typical values of k are much larger than the other macroscopic parameters. This leads to the approximation of solutions by asymptotic expressions.

The most direct way to construct an asymptotic solution of (5.12) is to choose a leading term having the form of a plane wave

$$u(x) \approx e^{i\kappa \psi(x)} Z(x). \tag{5.13}$$

Solutions to (5.12) will be approximated asymptotically by expanding Z as a formal (*i.e.*, possibly divergent) series in positive integral powers of k^{-1}. Here we neglect all but first-order terms in Z. We have

$$\nabla u = i\kappa \nabla \psi e^{i\kappa \psi} Z + e^{i\kappa \psi} \nabla Z,$$

and

$$-\kappa^2 e^{i\kappa \psi z} = -\kappa^2 u = \Delta u = \nabla \cdot \nabla u$$

$$= i\kappa \Delta \psi e^{i\kappa \psi} Z - \kappa^2 |\nabla \psi|^2 e^{i\kappa \psi} Z + 2i\kappa e^{i\kappa \psi} \nabla \psi \nabla Z + e^{i\kappa \psi} \Delta Z.$$

Neglecting the higher-order terms in Z, we take ΔZ to be zero. Dividing through by $\kappa \exp[i\kappa \psi]$, we obtain in place of (5.12) the complex equation

$$-\kappa Z = i\Delta \psi Z - \kappa |\nabla \psi|^2 Z + 2i\nabla \psi \cdot \nabla Z. \tag{5.14}$$

In order for (5.13) to satisfy (5.12), we require that the real and imaginary parts of (5.14) each vanish, that is, that

$$|\nabla \psi|^2 = 1 \tag{5.15}$$

and that

$$\Delta \psi Z + 2\nabla \psi \cdot \nabla Z = 0. \tag{5.16}$$

These are, respectively, the *eikonal* and *transport* equations. Because Z is proportional to the square root of the local ray density, (5.15) and (5.16) can be used to solve for both Z and the *phase* ψ in terms of the rays (see [13]). However, at *caustics* the rays coalesce and (5.13) no longer represents a solution, even in its qualitative behavior.

Remarks. *i)* The planar eikonal equation

$$\psi_x^2 + \psi_y^2 = 1 \tag{5.17}$$

has a simple geometric interpretation [2]. Denote by $\mathscr{D}$ a region bounded by a convex closed planar curve Σ having coordinates (x, y). It follows from the geometric-variational meaning of the gradient that any smooth function ψ defined in the complement of $\mathscr{D}$ and taking a value at each point equal to the distance from that point to Σ must satisfy (5.17).

ii) If we replace κ in (5.13) by k, then we obtain the eikonal equation in the alternative form

$$|\nabla \psi|^2 = v^2. \tag{5.18}$$

In fact the mathematical operations are generally analogous for any constant refractive index v, and unless the notation explicitly suggests otherwise (as in Sect. 5.2), the reader should assume that v is equal to 1.

5.1   A Uniform Asymptotic Expansion

An asymptotic expansion which retains its validity on both sides of a smooth convex caustic was introduced independently by Kravtsov [14] and Ludwig [16] in the mid-1960s. As usual, we restrict our attention to the plane. Let

$$u(x, y) = e^{ik\theta(x,y)} \left\{ g_0(x, y) A \left(k^{2/3} \rho(x, y) \right) + \frac{g_1(x, y)}{i k^{1/3}} A' \left(k^{2/3} \rho(x, y) \right) \right\}, \tag{5.19}$$

where ρ, θ, g_0, and g_1 are functions which do not depend on k and which are to be determined with the solution; the function A is the *Airy function*

$$A(k) = \frac{1}{2\pi i} \int_{T(\gamma')} \exp \left[\frac{u^3}{3} - ku \right] du \tag{5.20}$$

with typical initial conditions

$$A(0) = \frac{3^{-2/3}}{\Gamma(2/3)}$$

and

$$A'(0) = -\frac{3^{-1/3}}{\Gamma(1/3)},$$

where $\Gamma(\ )$ is the gamma function. (The initial conditions on A can be justified by computing its Maclaurin expansion.)

Substituting (5.19) into (5.12) and collecting terms yields

$$
e^{-ik\theta}\left(\Delta u + k^2 u\right) \approx -k^2 A g_0 \left[(\nabla\theta)^2 - \rho\,(\nabla\rho)^2 - 1\right]
$$
$$
-ik^{5/3} A' g_1 \left[(\nabla\theta)^2 - \rho\,(\nabla\rho)^2 - 1\right] + ik^{5/3} A' g_0 \left[2\nabla\theta \cdot \nabla\rho\right]
$$

$$(5.21)$$

plus terms of lower order in k. The right-hand side of this expression vanishes provided

$$(\nabla\theta)^2 - \rho\,(\nabla\rho)^2 - 1 = 0 \tag{5.22}$$

and

$$2\nabla\theta \cdot \nabla\rho = 0. \tag{5.23}$$

These equations will be shown to be equivalent to the eikonal equation (5.15). If one includes terms of the next highest order in k on the right-hand side of (5.21) and requires those terms to vanish, one obtains a system which can be shown to be analogous to the transport equation (5.16); see Sect. 1 of [16] for details.

Note that the oscillatory behavior of the Airy function occurs on the region where ρ is negative and the rapid quenching occurs on the region on which ρ is positive. (The opposite behavior is identified in [16], because that expansion replaces our $A(k)$ by $A(-k)$. The reason for our sign convention will become clear in Sect. 5.4, in which we take advantage of the fact that, in the present notation, the elliptic region of an associated differential equation surrounds its hyperbolic region.)

Multiplying (5.23) by $\pm\sqrt{\rho}$ on the region of positive ρ and adding the result to (5.22) yields

$$
1 = |\nabla\theta|^2 \pm 2\sqrt{\rho}\nabla\theta \cdot \nabla\rho + \left(\sqrt{\rho}\nabla\rho\right)^2 = \left(\nabla\theta \pm \sqrt{\rho}\nabla\rho\right)^2. \tag{5.24}
$$

Define

$$
\varphi^{\pm} = \theta \pm \frac{2}{3}\rho^{3/2}.
$$

Then we obtain from (5.22)–(5.24) a pair of eikonal equations having the form

$$
\left|\nabla\varphi^{\pm}\right|^2 = 1. \tag{5.25}
$$

5.2   The Complex Eikonal Equation

This section closely follow the analysis of [17], Sect. 1.

The eikonal equation for a real-valued function $\psi\,(x, y)$ becomes a first-order system if ψ is replaced by the complex function

$$
f\,(x, y) = u\,(x, y) + iv\,(x, y).
$$

Writing the eikonal equation (5.18) for f in terms of its real and imaginary parts, we obtain

$$u_x^2 + u_y^2 - v_x^2 - v_y^2 + v^2 = 0, \tag{5.26}$$

$$u_x v_x + u_y v_y = 0. \tag{5.27}$$

The complex-valued eikonal equation has acquired importance in optics in connection with the method of *evanescent wave tracking* [7, 8].

In terms of its geometry, (5.26) asserts that the squared length of ∇v differs from the squared length of ∇u by the number v^2. Equation (5.27) conveys the geometric information that ∇u and ∇v are orthogonal. This implies – as in the case of (5.23) – that the level curves of one function are the curves of steepest descent of the other function [9]. If

$$u_x^2 + u_y^2 > 0, \tag{5.28}$$

then v_x and v_y cannot both vanish, by (5.26). In that case, these geometric relations can be expressed, in the language of proportions, by the assertion that either

$$\begin{bmatrix} v_x \\ v_y \end{bmatrix} : \sqrt{v_x^2 + v_y^2} = \pm \begin{bmatrix} -u_y \\ u_x \end{bmatrix} : \sqrt{u_x^2 + u_y^2}, \tag{5.29}$$

or

$$u_x = u_y = 0. \tag{5.30}$$

These relations imply a coupled system of scalar equations:

$$\frac{v_x}{\sqrt{v_x^2 + v_y^2}} = \pm \frac{-u_y}{\sqrt{u_x^2 + u_y^2}}.$$

Cross multiplying,

$$v_x \sqrt{u_x^2 + u_y^2} = \mp u_y \sqrt{v_x^2 + v_y^2}.$$

Using (5.26),

$$v_x \sqrt{u_x^2 + u_y^2} = \mp u_y \sqrt{v^2 + u_x^2 + u_y^2},$$

or

$$v_x = \mp u_y \sqrt{\frac{v^2}{u_x^2 + u_y^2} + 1}. \tag{5.31}$$

The scalar equation

$$\frac{v_y}{\sqrt{v_x^2 + v_y^2}} = \pm \frac{u_x}{\sqrt{u_x^2 + u_y^2}},$$

which also follows from (5.29), implies by analogous operations the equation

$$v_y = \pm u_x \sqrt{\frac{v^2}{u_x^2 + u_y^2} + 1}. \tag{5.32}$$

The coupled system (5.31), (5.32) can be written as a vector equation of the form

$$\begin{bmatrix} v_x \\ v_y \end{bmatrix} = \pm \sqrt{1 + \frac{v^2}{u_x^2 + u_y^2}} \begin{bmatrix} -u_y \\ u_x \end{bmatrix}, \tag{5.33}$$

or as a single exact equation for 1-forms,

$$dv = \pm \sqrt{1 + \frac{v^2}{u_x^2 + u_y^2}} \left(-u_y dx + u_x dy \right). \tag{5.34}$$

This implies the local existence of a solution to the divergence-form equation

$$\frac{\partial}{\partial x} \left(\sqrt{1 + \frac{v^2}{u_x^2 + u_y^2}} u_x \right) + \frac{\partial}{\partial y} \left(\sqrt{1 + \frac{v^2}{u_x^2 + u_y^2}} u_y \right) = 0, \tag{5.35}$$

whenever condition (5.28) is satisfied.

In addition to being solvable for v_x and v_y as in (5.31) and (5.32), the equations (5.26), (5.27), and (5.29) can also be solved for u_x and u_y. If either

$$v_x^2 + v_y^2 = v^2 \tag{5.36}$$

or

$$v_x^2 + v_y^2 > v^2, \tag{5.37}$$

one obtains, by analogous arguments to those applied to v_x and v_y, the vector equation

$$\begin{bmatrix} u_x \\ u_y \end{bmatrix} = \mp \sqrt{1 - \frac{v^2}{v_x^2 + v_y^2}} \begin{bmatrix} -v_y \\ v_x \end{bmatrix} \tag{5.38}$$

and the equation for 1-forms,

$$du = \mp \sqrt{1 - \frac{v^2}{v_x^2 + v_y^2}} \left(-v_y dx + v_x dy \right). \tag{5.39}$$

Equation (5.39) is exact if either (5.36) holds, or if (5.37) holds as well as the divergence-form equation

$$\frac{\partial}{\partial x} \left(\sqrt{1 - \frac{v^2}{v_x^2 + v_y^2}} v_x \right) + \frac{\partial}{\partial y} \left(\sqrt{1 - \frac{v^2}{v_x^2 + v_y^2}} v_y \right) = 0. \tag{5.40}$$

We reproduce four important remarks from [17]:

i) Equations (5.35) and (5.40) can be associated to a Bäcklund transformation $u \to v$ and its inverse $v \to u$ defined by (5.33) and (5.38), in the sense that the map $u \to v$ defined by (5.33) and the inverse map $v \to u$ defined by (5.38) convert any solution of (5.35) satisfying (5.28) into a solution of (5.40) satisfying (5.37).

(Historically, the term *Bäcklund transformation* has been defined in many ways. A source for the classical theory is the monograph [27]. Here and below we use the term in the general sense of a function that maps a solution a of a differential equation A into a solution b of a differential equation B and vice-versa, where B may equal A but b will not equal a.)

ii) This is not the only Bäcklund transformation implied by the above construction. For example, in the limiting case $v(x, y) = 0$, we obtain a different Bäcklund transformation – the familiar one taking weighted Cauchy–Riemann equations (in two copies, (5.33) and (5.38)) into a weighted Laplace's equation (also in two copies, (5.35) and (5.40)).

iii) Every real-valued solution of the real-valued eikonal equation (5.18) satisfies (5.40).

iv) Many special solutions exist to equations having the general structure of (5.35) and (5.40). For example, the function

$$u(r) = a \left\{ \sqrt{1 - \frac{r^2}{a^2}} + \log \left(r / \left(|a| + \sqrt{a^2 - r^2} \right) \right) \right\} + b, \qquad (5.41)$$

where $r = \sqrt{x^2 + y^2}$ and a and b are constants with $a \neq 0$, satisfies a radial variant of (5.35) except on the set $\{r \,|\, r\,du/dr = 0\}$. The graph of $u(r)$ has a geometric interpretation as a *pseudosphere* (Fig. 5.1): a complete, simply connected surface having curvature equal to -1 times the curvature of a sphere having the same volume; c.f. [17], Sect. 2.3.

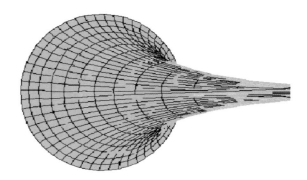

Fig. 5.1 The pseudosphere obtained by rotating the curve $y = u(x)$, where u is given by (5.41) with $a > 0$ and $b = 0$, about the positive x-axis

5.3   The Hodograph Map

The system (5.22), (5.23) will be satisfied by any solution of the eikonal-like system $\nabla \rho = 0$, $|\nabla \theta|^2 = 1$. An alternative choice is to replace (5.22), (5.23) by the scalar equation [18]

$$\left(|\nabla \theta|^4 - \theta_y^2 \right) \theta_{xx} + 2\theta_x \theta_y \theta_{xy} + \left(|\nabla \theta|^4 - \theta_x^2 \right) \theta_{yy} = 0. \tag{5.42}$$

(See also [17, 19, 20, 31] for careful discussions of (5.42) and its context.)

In order to find the domain on which the characteristic form associated with the operator of (5.42) has real roots, we convert that equation to a first-order system having the form

$$A^1 \Theta_x + A^2 \Theta_y = 0, \tag{5.43}$$

where the zero on the right-hand side denotes the 2×2 zero matrix; Θ is the vector (θ_x, θ_y);

$$A^1 = \begin{pmatrix} |\nabla \theta|^4 - \theta_y^2 & 0 \\ 0 & -1 \end{pmatrix}; \tag{5.44}$$

$$A^2 = \begin{pmatrix} 2\theta_x \theta_y & |\nabla \theta|^4 - \theta_x^2 \\ 1 & 0 \end{pmatrix}. \tag{5.45}$$

Proceeding as in Sect. 2.1, we associate a quadratic form $\tilde{Q}$ to (5.43) and evaluate the determinant

$$|\tilde{Q}| = \left| A^1 - \lambda A^2 \right| = -\left[\lambda^2 \left(|\nabla \theta|^4 - \theta_x^2 \right) - 2\lambda \theta_x \theta_y + |\nabla \theta|^4 - \theta_y^2 \right]. \tag{5.46}$$

We find that $\tilde{Q}$ has real roots only on each point of the closed disc $\theta_x^2 + \theta_y^2 \leq 1$. We conclude that the equation is elliptic on the $\mathbf{R}^2$-complement that disc. Due to the quasilinearity of the differential operator, we cannot locate the region of ellipticity in terms of x and y without explicitly solving the equation.

Thus it is necessary to linearize the equation if we are to know even the region of the xy-plane on which it is elliptic.

If a system of equations has the form

$$\begin{bmatrix} a_{11} & a_{12} \\ a_{21} & a_{22} \end{bmatrix} \frac{\partial}{\partial x} \begin{pmatrix} p \\ q \end{pmatrix} + \begin{bmatrix} b_{11} & b_{12} \\ b_{21} & b_{22} \end{bmatrix} \frac{\partial}{\partial y} \begin{pmatrix} p \\ q \end{pmatrix} = \begin{pmatrix} 0 \\ 0 \end{pmatrix}, \tag{5.47}$$

where the entries of the coefficient matrices depend only on p and q, then the coordinate transformation $(x, y) \rightarrow (p, q)$ takes (5.47) into a linear equation having the form

$$\begin{bmatrix} b_{12} & -a_{12} \\ b_{22} & -a_{22} \end{bmatrix} \frac{\partial}{\partial p} \begin{pmatrix} x \\ y \end{pmatrix} + \begin{bmatrix} -b_{11} & a_{11} \\ -b_{21} & a_{21} \end{bmatrix} \frac{\partial}{\partial q} \begin{pmatrix} x \\ y \end{pmatrix} = \begin{pmatrix} 0 \\ 0 \end{pmatrix}.$$

This transformation is called a *hodograph map*. It fails at any point for which its Jacobian $J = p_x q_y - p_y q_x$ vanishes.

This map is easy to understand as an inversion of the differential forms

$$dp = p_x dx + p_y dy$$

and

$$dq = q_x dx + q_y dy.$$

Written in matrix form, this is the system

$$\begin{pmatrix} dp \\ dq \end{pmatrix} = \begin{pmatrix} p_x & p_y \\ q_x & q_y \end{pmatrix} \begin{pmatrix} dx \\ dy \end{pmatrix},$$

which has the inverse

$$\begin{pmatrix} dx \\ dy \end{pmatrix} = \frac{1}{J} \begin{pmatrix} q_y & -p_y \\ -q_x & p_x \end{pmatrix} \begin{pmatrix} dp \\ dq \end{pmatrix}$$

provided J is nonvanishing.

The coordinate systems (p,q) and (x,y) are related by the *Lengendre transformation*

$$V(p,q) = xp + yq - v(x,y), \tag{5.48}$$

where in this case

$$(x,y) = \left(\frac{\partial V}{\partial p}, \frac{\partial V}{\partial q} \right)$$

and

$$(p,q) = \left(\frac{\partial v}{\partial x}, \frac{\partial v}{\partial y} \right).$$

Making the substitutions $p = \theta_x$, $q = \theta_y$ in (5.43)–(5.45), we obtain

$$\left[\left(p^2 + q^2 \right)^2 - q^2 \right] p_x + 2pq p_y + \left[\left(p^2 + q^2 \right)^2 - p^2 \right] q_y = 0, \tag{5.49}$$

$$p_y - q_x = 0. \tag{5.50}$$

Applying a hodograph transformation to (5.49), (5.50) yields

$$\left[\left(p^2 + q^2 \right)^2 - p^2 \right] x_p - 2pq x_q + \left[\left(p^2 + q^2 \right)^2 - q^2 \right] y_q = 0, \tag{5.51}$$

$$x_q - y_p = 0. \tag{5.52}$$

This system is, in turn, equivalent to the (now linear) scalar equation [18]

$$\left[\left(p^2 + q^2\right)^2 - p^2\right] V_{pp} - 2pqV_{pq} + \left[\left(p^2 + q^2\right)^2 - q^2\right] V_{qq} = 0, \qquad (5.53)$$

provided there is a continuously differentiable scalar function $V(x, y)$ for which $V_p = x$ and $V_q = y$.

Using the methods previously applied to (5.42), we find that (5.53) is elliptic on the $\mathbf{R}^2$-complement of the closed unit disc centered at the origin of coordinates in the hodograph plane. On this region the equation is locally equivalent to the potential equation and solutions having highly oscillatory boundary data will exhibit damped oscillations outside the unit disc. On the interior of the disc, (5.53) is locally equivalent to the wave equation and solutions propagate along the characteristic curves (wavefronts) of the solution, focussing at the center of the disc; c.f. Fig. 1 of [18].

Whereas the hodograph mapping replaces a quasilinear equation by a linear one, it tends to replace linear boundary conditions by nonlinear ones. In general this restricts the boundary conditions that can be applied in this method to relatively simple examples. For instance, the vanishing of a solution on a circle in the hodograph plane pulls back, under a hodograph transformation, to a constant value of the solution on the corresponding circle in the physical plane.

Moreover, the hodograph transformation may be singular. Nevertheless, it can be shown that in our case any singular set on the elliptic region of the equations must consist of isolated points.

Proposition 5.1. *Let (u, v) satisfy the equations*

$$\left[f(u, v) - v^2\right] u_x + uv\left(v_x + u_y\right) + \left[f(u, v) - u^2\right] v_y = 0, \qquad (5.54)$$

$$u_y - v_x = 0, \qquad (5.55)$$

for

$$f(u, v) \equiv \left(u^2 + v^2\right)^2. \qquad (5.56)$$

Then the hodograph mapping

$$\widetilde{h} : (x(u, v), y(u, v)) \rightarrow (u(x, y), v(x, y))$$

and its inverse can only be singular at isolated points in the elliptic region unless the images of $\widetilde{h}$ and its inverse are constants.

Proof. Denote by J the Jacobian of $\widetilde{h}$ and by j the Jacobian of its inverse mapping. The elliptic region of (5.54), (5.55) is defined by the inequality $u^2 + v^2 < f$. The hodograph image of that system has the form

$$\left(f - u^2\right) x_u - 2uvx_v + \left(f - v^2\right) y_v = 0, \qquad (5.57)$$

$$x_v = y_u. \qquad (5.58)$$

Using (5.58),

$$j = x_u y_v - x_v y_u = x_u y_v - x_v^2.$$

Now solving for x_u in (5.57) and simplifying, we obtain

$$j = \frac{\left[(f - u^2) x_u - uvx_v \right]^2 + f\left(f - u^2 - v^2 \right) x_v^2}{-(f - u^2)(f - v^2)}. \tag{5.59}$$

Given that f is nonnegative, we find that if $u^2 + v^2 < f$, then j is negative unless $x_u = x_v = 0$. But x_v can vanish only if y_u vanishes as well, because of (5.58). So the only possibly nonvanishing partial derivative, when $j = 0$, is y_v. But that possibility is excluded, given the vanishing of x_u and x_v, by (5.57). So the vanishing of j implies that x and y must be constants. Otherwise, x and y satisfy a linear, elliptic partial differential equation having analytic coefficients. The theory of analytic functions now implies the assertion of the theorem. A similar argument can be constructed for the inverse of j, the Jacobian

$$J = u_x y_y - v_x u_y = u_x v_y - v_x^2.$$

We find that J does not vanish for $u^2 + v^2 < f$ unless $u_x = v_x = 0$. But then $u_y = 0$ by condition (5.55). Now $v_y = 0$ by (5.54). So either u and v are constants, or $J \neq 0$ and the equation is analytic in the elliptic region. This completes the proof.   □

There are interesting choices for f other than (5.56). For example, if $f \equiv 1$, then the hodograph image of (5.54), (5.55) is identical to the Hodge equations on the projective disc (Sect. 6.1). A version of this result also holds for the continuity equation for an ideal gas; c.f. [5], (12.4.12) and (12.4.13). The proof of the proposition extends immediately to any $f(u, v)$ which is an analytic function of u and v.

5.4 A Closed Dirichlet Problem Which Is Weakly Well-Posed

The existence of solutions to the Dirichlet problem for (5.53) has been shown by Magnanini and Talenti. These solutions have a point singularity at the origin of coordinates in the hodograph plane.

Theorem 5.1 (Magnanini–Talenti [18]). *Let D_R, $R > 1$, be a disc of radius R centered at the origin of coordinates in $\mathbf{R}^2$ and let Ω be a subset of D_R that has positive distance from both the boundary and the center of D_R. Suppose that $f \in L^2[-\pi, \pi]$ is a given function. Choose polar coordinates (r, θ), where $0 < r < \infty$, $-\pi < \theta \leq \pi$. Then there is a unique function V lying in $H^{1,2}(\Omega)$, which weakly satisfies (5.53) in the polar form*

$$\left(r^2 - 1\right) V_{rr} + rV_r + V_{\theta\theta} = 0 \tag{5.60}$$

in D_R and is a C^∞ solution of (5.60) in the punctured disc D_r, where $0 < r < R$. Moreover, $V = f$ on ∂D in the sense that

$$\lim_{r \uparrow R} \int_{-\pi}^{\pi} |V\left(r \cdot e^{i\theta}\right) - f(\theta)|^2 d\theta = 0.$$

Finally, V has an explicit representation in the form of a series

$$V\left(re^{i\theta}\right) = \frac{a_0}{2} + \sum_{k=1}^{\infty} \frac{T_k(r)}{T_k(R)} \cdot \{a_k \cos(k\theta) + b_k \sin(k\theta)\}, \tag{5.61}$$

where

$$a_k = \pi^{-1} \int_{-\pi}^{\pi} f(\theta) \cos(k\theta) \, d\theta;$$

$$b_k = \pi^{-1} \int_{-\pi}^{\pi} f(\theta) \sin(k\theta) \, d\theta;$$

$$T_k(r) = \begin{cases} \cos(k \arccos r) & \text{if } r \leq 1, \\ \frac{1}{2}\left[\left(r + \sqrt{r^2 - 1}\right)^k + \left(r - \sqrt{r^2 - 1}\right)^k\right] & \text{if } r > 1. \end{cases}$$

Proof. We first demonstrate that any weak solution that exists is unique. Indeed, Parseval's Identity implies that on Ω, any solution V can be represented by its Fourier coefficients

$$C_k(r) = \frac{1}{\pi} \int_{-\pi}^{\pi} V\left(r \cdot e^{ik\theta}\right) \tilde{A}(k\theta) \, d\theta,$$

where $k \in \mathbf{N} \cup \{0\}$, $\tilde{A}$ is either the sine or the cosine function, and by hypothesis

$$\int_{\Omega} \left[C_k'(r)^2 + \frac{1}{r^2} C_k(r)^2\right] r dr < \infty.$$

That is to say, these objects are sufficiently integrable to be themselves a weak solution. Writing the weak form of (5.60) in terms of $C_k(r)$, we obtain the expression

$$\int_0^R \{(r^2 - 1) C_k'(r)\psi'(r) + k^2 C_k(r)\psi(r)\} \frac{dr}{\sqrt{|r^2 - 1|}} = 0 \tag{5.62}$$

$\forall \psi \in C_0^\infty((0, R))$. Integrating by parts in the highest-order term, we have for $r \neq 1$,

$$\int_0^R \left(r^2 - 1\right) C'(r) \psi'(r) \left(|r^2 - 1|\right)^{-1/2} dr$$

$$= \int_0^R \frac{d}{dr} \left(\frac{r^2 - 1}{\sqrt{|r^2 - 1|}} C'(r) \psi(r)\right) dr - \int_0^R 2r C'(r) \psi(r) \frac{dr}{\sqrt{|r^2 - 1|}}$$

$$- \int_0^R \left(r^2 - 1\right) C'(r) \psi(r) \left(-\frac{1}{2}\right) \left(|r^2 - 1|\right)^{-3/2} 2r\, dr$$

$$- \int_0^r \left(r^2 - 1\right) C''(r) \psi(r) \frac{dr}{\sqrt{|r^2 - 1|}}$$

$$= \int_0^R \frac{d}{dr} \left(\frac{r^2 - 1}{\sqrt{|r^2 - 1|}} C'(r) \psi(r)\right) dr$$

$$+ \int_0^R \frac{(-2r + r) C'(r) - (r^2 - 1) C''(r)}{\sqrt{|r^2 - 1|}} \psi(r)\, dr$$

$$= - \int_0^R \frac{(r^2 - 1) C''(r) + r C'(r)}{\sqrt{|r^2 - 1|}} \psi(r)\, dr,$$

where we have used the compact support of ψ to set the boundary term equal to zero. Adding the zeroth-order term in (5.62), we find that the coefficients $C_k(r)$ satisfy the *Chebyshev differential equation*

$$\left(r^2 - 1\right) C_k''(r) + r C_k'(r) - k^2 C_k(r) = 0, \ r \neq 1,$$

where

$$\lim_{r \to 1^-} \frac{r^2 - 1}{\sqrt{|r^2 - 1|}} C_k'(r) = \lim_{r \to 1^+} \frac{r^2 - 1}{\sqrt{|r^2 - 1|}} C_k'(r).$$

Any solution of this equation is proportional to a Chebyshev polynomial of order k. We conclude that V has the form

$$V\left(r \cdot e^{i\theta}\right) = \frac{\alpha_0}{2} + \sum_{k=1}^\infty T_k(r) \cdot \{\alpha_k \cos(k\theta) + \beta_k \sin(k\theta)\}, \tag{5.63}$$

where $T_k(r)$ is the kth Chebyshev polynomial. Letting r tend to R, we have

$$\alpha_k = \frac{a_k}{T_k(R)}$$

and

$$\beta_k(r) = \frac{b_k}{T_k(R)}.$$

So any V that exists and has the properties claimed must be given by (5.61), which proves uniqueness.

In order to prove existence we need only show that the function V given by (5.61), which is a classical solution in the punctured disc, is a weak solution in the entire disc.

Consider the energy functional

$$E = \int \int_{D_R} \left[(r^2 - 1) \left(\frac{\partial V}{\partial r} \right)^2 + \left(\frac{\partial V}{\partial \theta} \right)^2 \right] |r^2 - 1|^{-1/2} \, dr d\theta.$$

The variational equations of E take the form

$$\frac{\partial}{\partial r} \left[(r^2 - 1) |r^2 - 1|^{-1/2} \cdot \frac{\partial V}{\partial r} \right] + |r^2 - 1|^{-1/2} \frac{\partial^2 V}{\partial \theta^2} = 0, \qquad (5.64)$$

which is equivalent to (5.60). The weak form of (5.64) can be written

$$I = \lim_{\varepsilon \to 0} \int_{-\pi}^{\pi} \int_{\varepsilon}^{R} \left[(r^2 - 1) \frac{\partial V}{\partial r} \frac{\partial Z}{\partial r} + \frac{\partial V}{\partial \theta} \frac{\partial Z}{\partial \theta} \right] |r^2 - 1|^{-1/2} \, dr d\theta$$

for arbitrary $Z \in C_0^\infty (D_R)$. Integrating I by parts, we obtain

$$I = \lim_{\varepsilon \to 0} \int_{-\pi}^{\pi} \int_{\varepsilon}^{R} \frac{\partial}{\partial r} \left[(r^2 - 1) |r^2 - 1|^{-1/2} \cdot \frac{\partial V}{\partial r} Z \right] dr d\theta$$

$$- \lim_{\varepsilon \to 0} \int_{-\pi}^{\pi} \int_{\varepsilon}^{R} \frac{\partial}{\partial r} \left[(r^2 - 1) |r^2 - 1|^{-1/2} \frac{\partial V}{\partial r} \right] Z \, dr d\theta$$

$$+ \lim_{\varepsilon \to 0} \int_{-\pi}^{\pi} \int_{\varepsilon}^{R} \left[\frac{\partial}{\partial \theta} \left(\frac{\partial V}{\partial \theta} Z \right) - \left(\frac{\partial^2 V}{\partial \theta^2} \right) Z \right] |r^2 - 1|^{-1/2} \, dr d\theta.$$

Because Z has compact support in D_R and

$$\lim_{\varepsilon \to 0} (\varepsilon^2 - 1) |\varepsilon - 1|^{-1/2} = -1,$$

we have

$$I = \lim_{\varepsilon \to 0} \int_{-\pi}^{\pi} \frac{\partial V}{\partial r} (\varepsilon \cdot e^{i\theta}) Z (\varepsilon \cdot e^{i\theta}) \, d\theta$$

$$- \lim_{\varepsilon \to 0} \int_{-\pi}^{\pi} \int_{\varepsilon}^{R} \left\{ \frac{\partial}{\partial r} \left[(r^2 - 1) |r^2 - 1|^{-1/2} \frac{\partial V}{\partial r} \right] + \frac{\partial^2 V}{\partial \theta^2} |r^2 - 1|^{-1/2} \right\} Z \, dr d\theta$$

$$+ \lim_{\varepsilon \to 0} \int_{\varepsilon}^{R} \left(\frac{\partial V}{\partial \theta} (r, \theta) \cdot Z (r, \theta) \right)_{\theta = -\pi}^{\theta = \pi} |r^2 - 1|^{-1/2} \, dr$$

$$\equiv i_1 + i_2 + i_3.$$

Now i_2 vanishes because V is a smooth solution of (5.64) on $D_R \setminus \{0\}$; i_3 vanishes by the properties of Z and (5.63); we only need to estimate i_1. Because Z is Lipschitz continuous,

$$|i_1| \le \left| \int_{-\pi}^{\pi} \frac{\partial V}{\partial r} \left(\varepsilon \cdot e^{i\theta} \right) Z \left(\varepsilon \cdot e^{i\theta} \right) d\theta \right|$$

$$= \left| \int_{-\pi}^{\pi} \frac{\partial V}{\partial r} \left(\varepsilon \cdot e^{i\theta} \right) \left[Z \left(\varepsilon \cdot e^{i\theta} \right) - Z(0) + Z(0) \right] d\theta \right|$$

$$\le \left| \int_{-\pi}^{\pi} \frac{\partial V}{\partial r} \left(\varepsilon \cdot e^{i\theta} \right) \left[Z \left(\varepsilon \cdot e^{i\theta} \right) - Z(0) \right] d\theta \right| + |Z(0)| \int_{-\pi}^{\pi} \frac{\partial V}{\partial r} \left(\varepsilon \cdot e^{i\theta} \right) d\theta.$$

The last density on the right integrates to zero. Thus we have

$$|i_1| = \left| \int_{-\pi}^{\pi} \frac{\partial V}{\partial r} \left(\varepsilon \cdot e^{i\theta} \right) \left[Z \left(\varepsilon \cdot e^{i\theta} \right) - Z(0) \right] d\theta \right|$$

$$\le C \cdot \varepsilon \cdot \left\{ \int_{-\pi}^{\pi} \left[\frac{\partial V}{\partial r} \left(\varepsilon \cdot e^{i\theta} \right) \right]^2 d\theta \right\}^{1/2} \le \tilde{C} \cdot \varepsilon,$$

where C and $\tilde{C}$ are finite, positive constants that are bounded in the limit as ε tends to zero. Taking that limit completes the proof.   □

5.5 A Class of Strongly Well-Posed Boundary Value Problems

Consider second-order equations having the form

$$[\mathscr{K}(r)u_r]_r + u_{\theta\theta} + \tilde{k} u_\theta = f(r, \theta), \tag{5.65}$$

where $\mathscr{K}'(r) > 0$; $\tilde{k}$ is a nonzero constant; $\mathscr{K}(r) < 0$ for $0 \le r < r_{crit}$ and $\mathscr{K}(r) > 0$ for $r_{crit} < r \le R$. Rather than study this equation directly, we consider the associated system

$$Lw = A^1 w_r + A^2 w_\theta + Bw = F, \tag{5.66}$$

where L is a first-order operator, $w = (w_1(r, \theta), w_2(r, \theta))$, $F = (f, 0)$,

$$A^1 = \begin{pmatrix} \mathscr{K}(r) & 0 \\ 0 & -1 \end{pmatrix}, \quad A^2 = \begin{pmatrix} 0 & 1 \\ 1 & 0 \end{pmatrix}, \tag{5.67}$$

and

$$B = \begin{pmatrix} \mathcal{K}'(r) & \tilde{k} \\ 0 & 0 \end{pmatrix}. \tag{5.68}$$

We interpret r as a radial coordinate and θ as an angular coordinate, so that the system (5.66)–(5.68) is defined on a closed disc of radius R. The system is equivalent to (5.65) if the components of w are C^2 and $w_1 = u_r$, $w_2 = u_\theta$.

Theorem 5.2 ([25]). *Suppose that there is a positive constant v_0 such that $\mathcal{K}'(r) \geq v_0$. Let there be continuous functions $\sigma(\theta)$ and $\tau(\theta)$ such that the boundary condition*

$$\sigma(\theta)w_1 + \tau(\theta)w_2 = 0 \tag{5.69}$$

is satisfied on the boundary $r = R$, where the product $\sigma(\theta)\tau(\theta)$ is either strictly positive or strictly negative and has sign opposite to the sign of $\tilde{k}$. Then the boundary value problem (5.66)–(5.69), with $\mathcal{K}$ and $\tilde{k}$ as defined in (5.65), possesses a strong solution on the closed disc $\{(r, \theta) \mid 0 \leq r \leq R\}$ provided $|\mathcal{K}(0)|$ is sufficiently small.

Proof. Multiply the terms of (5.66) by the matrix

$$E = \begin{pmatrix} a & -c\mathcal{K}(r) \\ c & a \end{pmatrix},$$

where a and c are constants; the sign of c is chosen so that $\sigma\tau c < 0$ (so $c\tilde{k} > 0$); $a > 0$; and $|c|$ is large. The matrix E is nonsingular provided

$$\det E = a^2 + c^2 \mathcal{K}(r) \neq 0.$$

Because $\mathcal{K}(r)$ is continuous, increasing, and $\mathcal{K}(r_{crit}) = 0$, the invertibility condition for E becomes

$$\min_{r \in [0,R]} |\mathcal{K}(r)| < \frac{a^2}{c^2}.$$

This condition will be satisfied provided $|\mathcal{K}(0)|$ is sufficiently small.

The symmetric part κ^* of the matrix

$$\kappa = EB - (1/2)\left[(EA^1)_r + (EA^2)_\theta\right]$$

has determinant

$$\Delta = \frac{a\tilde{k}}{2}\left[c\mathcal{K}'(r) - \frac{a\tilde{k}}{2}\right] \geq \frac{a\tilde{k}}{2}\left[cv_0 - \frac{a\tilde{k}}{2}\right],$$

so the resulting system is symmetric positive provided $|c|$ is sufficiently large. The proof will be complete once we show that the boundary conditions are admissible. Proceeding as in [32], choose the outward-pointing normal vector to have the form $n = \mathscr{K}^{-1}(r)dr$. (We can always do this because the domain is a disc.)

Then on the boundary $r = R$,

$$\beta = \begin{pmatrix} a & c \\ c & -a\mathscr{K}^{-1}(R) \end{pmatrix}.$$

Choose

$$\beta_- = \frac{1}{\sigma^2 + \tau^2} \begin{pmatrix} \sigma\tau c + \sigma^2 a & \tau^2 c + \sigma\tau a \\ -\sigma\tau a\mathscr{K}^{-1}(R) + \sigma^2 c & -\tau^2 a\mathscr{K}^{-1}(R) + \sigma\tau c \end{pmatrix}$$

and $\beta_+ = \beta - \beta_-$. Then $\beta_- w = 0$, as (5.69) implies that $w_2 = -(\sigma/\tau)w_1$ on the circle $r = R$. Moreover,

$$\mu = \frac{1}{\sigma^2 + \tau^2} \begin{pmatrix} (\tau^2 - \sigma^2)a - 2\sigma\tau c & (\sigma^2 - \tau^2)c - 2\sigma\tau a \\ (\tau^2 - \sigma^2)c + 2\sigma\tau|a\mathscr{K}^{-1}(R)| & (\tau^2 - \sigma^2)a\mathscr{K}^{-1}(R) - 2\sigma\tau c \end{pmatrix},$$

implying that

$$\mu^* = \frac{1}{\sigma^2 + \tau^2} \begin{pmatrix} (\tau^2 - \sigma^2)a - 2\sigma\tau c & \sigma\tau a\left(\mathscr{K}^{-1}(R) - 1\right) \\ \sigma\tau a\left(\mathscr{K}^{-1}(R) - 1\right) & (\tau^2 - \sigma^2)a\mathscr{K}^{-1}(R) - 2\sigma\tau c \end{pmatrix}.$$

If $\sigma\tau < 0$, then $c > 0$; if $\sigma\tau > 0$, then $c < 0$. In either case the matrix μ^* will be non-negative provided $|c|$ is sufficiently large.

Now

$$\mathscr{R}_- = \frac{\sigma w_1 + \tau w_2}{\sigma^2 + \tau^2} \begin{pmatrix} \tau c + \sigma a \\ -\tau a\mathscr{K}^{-1}(R) + \sigma c \end{pmatrix}$$

and

$$\mathscr{R}_+ = \frac{\tau w_1 - \sigma w_2}{\sigma^2 + \tau^2} \begin{pmatrix} \tau a - \sigma c \\ \sigma a\mathscr{K}^{-1}(R) + \tau c \end{pmatrix};$$

so $\mathscr{R}_- \cap \mathscr{R}_+ = 0$, where $\mathscr{R}_\pm$ are defined as in Sect. 2.5. Because conditions are given on the entire boundary of the disc, the null space of β_- alone spans the range of ∂V.

The invertibility of E completes the proof of Theorem 5.2.   □

A similar result has been proven by Torre [32] for the equation

$$\frac{1}{r}(ru_r)_r + \left(\frac{1}{r^2} - \tilde{\omega}^2\right)u_{\theta\theta} = f(r, \theta), \tag{5.70}$$

where f is a sufficiently smooth function and $\tilde{\omega}$ is a constant. In [32], a "Sommerfeld" condition of the form (5.69) is imposed on functions σ, τ such that the product $\sigma\tau$ does not vanish on the outer boundary of an annulus and σ, τ satisfy the Dirichlet conditions $\sigma = 1$, $\tau = 0$ on the inner boundary. Because (5.70) is a helically reduced wave equation in $\mathbf{M}^{2,1}$, it provides a toy model for the reduction of the Einstein equations by a helical Killing field (c.f. Examples 1 and 4 of Sect. 6.4.5, below). In distinction to (5.65), which is of Keldysh type, (5.70) is of Tricomi type.

5.6 Hodge–Bäcklund Transformations

(This section, including Sect. 5.6.1, assumes familiarity with the geometry of differential forms, but can be skipped without loss of continuity.)

Equations (5.35) and (5.40) can each be written in the form of *nonlinear Hodge equations*

$$d * (\rho(Q)\omega) = 0, \tag{5.71}$$

$$d\omega = 0 \tag{5.72}$$

where $\omega \in \Lambda^1(T^*M)$; M is a two-dimensional Riemannian manifold; $* : \Lambda^p \to \Lambda^{2-p}$ is the Hodge involution; $Q = *(\omega \wedge *\omega) > 0$; $d : \Lambda^p \to \Lambda^{p+1}$ is the exterior derivative.

Either

$$\rho(|\omega|^2) = \sqrt{1 + \frac{v^2}{|\omega|^2}}, \tag{5.73}$$

which corresponds to (5.35), or

$$\hat{\rho}(|\xi|^2) = \sqrt{1 - \frac{v^2}{|\xi|^2}} \tag{5.74}$$

which corresponds to (5.40). In either case we assume Q to be nonvanishing, by analogy with the conditions (5.28), (5.30), (5.36), and (5.37).

Initially, take ρ as in (5.73). Provided the domain is simply connected, (5.72) implies that there is a 0-form u such that

$$\omega = du,$$

and a 0-form v such that

$$dv = \pm * \left(\sqrt{1 + \frac{v^2}{|\omega|^2}}\, \omega \right) = \pm * \left(\sqrt{|du|^2 + v^2}\, \frac{\omega}{|\omega|} \right). \tag{5.75}$$

Because the Hodge operator is an isometry – which is illustrated locally by (5.29), we have

$$|dv|^2 = |du|^2 + v^2.$$

Thus the transformation (5.75) yields an invariant form of (5.26). But in distinction to classical Bäcklund transformations of the eikonal equation, in the present case the Cauchy–Riemann equations are not satisfied. Rather,

$$u_x = \mp \rho(Q) v_y$$

and

$$u_y = \pm \rho(Q) v_x;$$

this is sufficient for the orthogonality condition (5.27) provided ρ does not cavitate.

Now take $\hat{\rho}$ as in (5.74). Arguing as before, we find that there is a 0-form $\tilde{u}$ such that $\omega = d\tilde{u}$, and a 0-form $\tilde{v}$ such that

$$d\tilde{v} = \pm * \left(\sqrt{Q - v^2} \frac{\omega}{|\omega|} \right).$$

This yields the relation

$$|d\tilde{v}|^2 = |d\tilde{u}|^2 - v^2.$$

Letting $\tilde{u} = \pm i u$ and $\tilde{v} = \pm i v$, we obtain a mapping taking solutions to (5.71), (5.72) and $\hat{\rho}$ satisfying (5.74) into solutions of the same system with ρ satisfying (5.73).

Motivated by these examples, we define a *Hodge–Bäcklund transformation* to be a map taking a solution a of a nonlinear Hodge equation having mass density ρ_A into a solution b of a nonlinear Hodge equation having mass density ρ_B and vice-versa, where B may equal A but b will not equal a. (A slightly more general definition, motivated by the discussion of Sect. 5.6.1, is given in Sect. 6.2.1 of [21].)

If ρ is taken to be constant, then (5.71), (5.72) are the Hodge–Kodaira equations for 1-forms, an invariant representation of Helmholtz's original vector formulation for the absence of sources and sinks; c.f. [10] and, e.g., [11].

The usefulness of expressing quasilinear field theories in the form (5.71), (5.72) goes well beyond the opportunity of using the Hodge–Bäcklund transformation rather than coordinate-based transformations. In the form (5.71), (5.72), the models of Sect. 2.7.2 inherit the rich geometry of harmonic forms and the topological methods for their analysis. In particular, one can derive existence theorems for the class as a whole (on the elliptic part of the domain and over a suitable underlying manifold) by an extension of the Hodge Decomposition Theorem [28, 29]; see also [12].

5.6.1 Hodge–Bäcklund Transformations in Fluid Dynamics and Geometry

Recall from item *i*) of Sect. 2.7.2 the mass density for the adiabatic and isentropic subsonic flow of an ideal fluid, (2.38), where $Q \in [0, 2/(\gamma + 1))$, is the squared flow speed; γ is the *adiabatic constant* – the ratio of specific heats for the medium. The adiabatic constant for air is 1.4. Choosing γ to be 2 one obtains, by a physical argument originally introduced for one space dimension [26], the mass density for shallow hydrodynamic flow in the tranquil regime [example *x*) of Sect. 2.7.2]. Choosing γ to be -1 (a physically impossible choice) leads to the density function for the minimal surface equation (2.39) (c.f. [15,29]). Flow governed by this density is called *Chaplygin flow*. Although the numbers -1 and 1.4 are not particularly close, this choice of mass density nevertheless has many attractive properties as an approximation for (2.38) ([4]; see also e.g., [6] and Chap. 5 of [3]).

The mathematical context of the nonlinear Hodge equations (5.71), (5.72) is clarified somewhat if they are generalized to the *nonlinear Hodge–Frobenius equations* ([21]; Sect. 4 of [24]; Sect. 6 of [23])

$$\delta\,[\rho(Q)\omega] = *(d\eta \wedge *\rho(Q)\omega)\,, \tag{5.76}$$

$$d\omega = d\eta \wedge \omega\,, \tag{5.77}$$

where $\delta : \Lambda^{p+1} \rightarrow \Lambda^p$ is the formal adjoint of d and $\eta \in \Lambda^0$ is given. Whereas a field ω satisfying (5.72) is conservative, a field satisfying (5.77) is only completely integrable. The nonzero right-hand side of (5.76) arises from variational conditions that are common in field theories; see Sect. 5.1 of [21].

The major differences between conservative fields and completely integrable ones are summarized in Tables 5.1 and 5.2:

Table 5.1 Some properties of conservative fields

Conservative field
$d\omega = 0$
Solutions lie in a cohomology class
Describes, e.g., irrotational flow
$\omega = d\varphi$ locally
Line integrals are independent of path

Table 5.2 Some properties of completely integrable fields

Completely integrable field
$d\omega = \Gamma \wedge \omega,\ \Gamma \in \Lambda^1$
Solutions lie in a closed ideal
Describes, e.g., rigid-body rotation
$\omega = e^\eta d\varphi$ locally, for $\Gamma = d\eta$
Integrals of $e^{-\eta}\omega$ are independent of path

We have the following result for solutions of (5.76), (5.77), which extends an argument introduced for the case $d\eta \equiv 0$ by Yang in [33]; see also [1] and Theorem 2.1 of [30].

Theorem 5.3 ([21]). *Let the 1-form ω satisfy (5.76) and (5.77), with ρ satisfying (2.39). Then there exists an $(n-1)$-form ξ with $|\xi| < 1$, satisfying equations analogous to (5.76) and (5.77), but with $d\eta$ replaced by $d\hat{\eta} = -d\eta$ and $\rho(Q)$ replaced by*

$$\hat{\rho}(|\xi|^2) \equiv \frac{1}{\sqrt{1 - |\xi|^2}} \, . \tag{5.78}$$

Proof. Equation (5.76) can be interpreted as the assertion that the $(n-1)$-form

$$\xi = * \left[\rho(|\omega|^2)\omega \right] = * \left[\frac{\omega}{\sqrt{1 + |\omega|^2}} \right] \tag{5.79}$$

satisfies

$$d\xi = d\hat{\eta} \wedge \xi, \tag{5.80}$$

that is, equation (5.77) with η replaced by $\hat{\eta} \equiv -\eta$.

Because the Hodge involution is an isometry, we have

$$|\xi|^2 = \frac{|\omega|^2}{1 + |\omega|^2}$$

or, equivalently,

$$1 - |\xi|^2 = \frac{1}{1 + |\omega|^2} \, . \tag{5.81}$$

Equation (5.81) implies that $|\xi|^2 < 1$, and also that

$$\rho(|\omega|^2)\hat{\rho}(|\xi|^2) = \frac{1}{\sqrt{1 + |\omega|^2}} \frac{1}{\sqrt{1 - |\xi|^2}} = \frac{1}{\sqrt{1 + |\omega|^2}} \cdot \sqrt{1 + |\omega|^2} = 1.$$

This, together with (5.79), yields directly

$$*\hat{\rho}(|\xi|^2)\xi = *^2 \hat{\rho}(|\xi|^2)\rho(|\omega|^2)\omega = (-1)^{n-1}\omega.$$

We conclude that

$$d * (\hat{\rho}(|\xi|^2)\xi) = (-1)^{n-1}d\omega = (-1)^{n-1}d\eta \wedge \omega$$
$$= d\eta \wedge * (\hat{\rho}(|\xi|^2)\xi) = -d\hat{\eta} \wedge * (\hat{\rho}(|\xi|^2)\xi) \, .$$

This is equivalent to equation (5.76) for the $(n-1)$-form ξ, where ρ has been replaced by $\hat{\rho}$ and η by $\hat{\eta}$, which completes the proof.   □

This argument carries over to any pairing of functions $\rho(|\omega|^2)$, $\hat{\rho}(|\xi|^2)$, as long as their product is 1.

References

1. Alías, L.J., Palmer, B: A duality result between the minimal surface equation and the maximal surface equation. An. Acad. Brasil. Ciênc. **73**, 161–164 (2001)
2. Arnold, V.I.: Lectures on Partial Differential Equations. Springer-Phasis, Berlin (2004)
3. Bers, L.: Mathematical Aspects of Subsonic and Transonic Gas Dynamics. Wiley, New York (1958)
4. Chaplygin, S.A.: On gas jets. Sci. Mem. Moscow Univ. Phys. Sec. **21**, 1–121 (1902) [Translation: NACA Tech. Mem. **1063** (1944)]
5. Chapman, C.J.: High Speed Flow. Cambridge University Press, Cambridge (2000)
6. Dinh, H., Carey, G.F.: Some results concerning approximation of regularized compressible flow. Int. J. Num. Meth. Fluids **5**, 299–302 (1985)
7. Felsen, L.B.: Evanescent waves. J. Opt. Soc. Amer. **66**, 751–760 (1976)
8. Felsen, L.B.: Complex spectra in high-frequency propagation and diffration. Proceedings of the 1985 International Symposium on Antennas and Propagation, Vol. 1, pp. 221–223. ISAP, Japan (1985)
9. Gerlach, U.H.: Linear Mathematics in Infinite Dimensions: Signals, Boundary Value Problems, and Special Functions, Lecture 46: The Method of Steepest Descent and Stationary Phase. http://www.math.osu.edu/~gerlach/math/BVtypset/node128.html.Cited2Aug2011
10. Helmholtz, H.: Über Integrale der hydrodynamischen Gleichungen, welche den Wirbelbewegungen entsprechen. J. Reine Angew. Math. **55**, 25–55 (1858)
11. Hodge, W.V.D.: A Dirichlet problem for harmonic functionals with applications to analytic varieties. Proc. London Math. Soc. **36**, 257–303 (1934)
12. Iwaniec, T., Scott, C., Stroffolini, B.: Nonlinear Hodge theory on manifolds with boundary. Annali Mat. Pura Appl. **177**, 37–115 (1999)
13. Keller, J.B.: Geometrical theory of diffraction. In: Graves, L.M. (ed.) Calculus of Variations and its Applications, Proceedings of Symposia in Applied Mathematics, vol. 8, pp. 27–52. McGraw-Hill, New York (1958)
14. Kravtsov, Yu.A.: A modification of the geometrical optics method [in Russian]. Radiofizika **7**, 664–673 (1964)
15. Kreyszig, E.: On the theory of minimal surfaces. In: Rassias, Th.M. (ed.) The Problem of Plateau: A Tribute to Jesse Douglas and Tibor Radó, pp. 138–164. World Scientific, Singapore (1992)
16. Ludwig, D.: Uniform asymptotic expansions at a caustic. Commun. Pure Appl. Math. **19**, 215–250 (1966)
17. Magnanini, R., Talenti, G.: On complex-valued solutions to a 2D eikonal equation. Part one: qualitative properties. Contemporary Math. **283**, 203–229 (1999)
18. Magnanini, R., Talenti, G.: Approaching a partial differential equation of mixed elliptic-hyperbolic type. In: Anikonov, Yu.E., Bukhageim, A.L., Kabanikhin, S.I., Romanov, V.G. (eds.) Ill-posed and Inverse Problems, pp. 263–276. VSP, Utrecht (2002)
19. Magnanini, R., Talenti, G.: On complex-valued solutions to a two-dimensional eikonal equation. II. Existence theorems. SIAM J. Math. Anal. **34**, 805–835 (2003)
20. Magnanini, R., Talenti, G.: On complex-valued solutions to a 2D eikonal equation. III. Analysis of a Bäcklund transformation. Appl. Anal. **85**, 249–276 (2006)
21. Marini, A., Otway, T.H.: Nonlinear Hodge-Frobenius equations and the Hodge-Bäcklund transformation. Proc. R. Soc. Edinburgh, Ser. A **140**, 787–819 (2010)

22. Morawetz, C.S.: Note on a maximum principle and a uniqueness theorem for an elliptic-hyperbolic equation. Proc. R. Soc. London, Ser. A **236**, 141–144 (1956)
23. Otway, T.H.: Nonlinear Hodge maps. J. Math. Phys. **41**, 5745–5766 (2000)
24. Otway, T.H.: Maps and fields with compressible density. Rend. Sem. Mat. Univ. Padova **111**, 133–159 (2004)
25. Otway, T.H.: Variational equations on mixed Riemannian-Lorentzian metrics. J. Geom. Phys. **58**, 1043–1061 (2008)
26. Riabouchinsky, D.: Sur l'analogie hydraulique des mouvements d'un fluide compressible. C. R. Academie des Sciences, Paris **195**, 998 (1932)
27. Rogers, C., Schief, W.K.: Bäcklund and Darboux Transformations, Geometry and Modern Applications of Soliton Theory. Cambridge University Press, Cambridge (2002)
28. Sibner, L.M., Sibner, R.J.: A nonlinear Hodge-de Rham theorem. Acta Math. **125**, 57–73 (1970)
29. Sibner, L.M., Sibner, R.J.: Nonlinear Hodge theory: Applications. Advances in Math. **31**, 1–15 (1979)
30. Sibner, L.M., Sibner, R.J., Yang, Y.: Generalized Bernstein property and gravitational strings in Born–Infeld theory. Nonlinearity **20**, 1193–1213 (2007)
31. Talenti, G.: Some equations of non-geometrical optics. In: Lupo, D., Pagani, C.D., Ruf. B. (eds.) Nonlinear Equations: Methods, Models and Applications, pp. 257–267. Birkhäuser, Basel (2003)
32. Torre, C.G.: The helically reduced wave equation as a symmetric positive system. J. Math. Phys. **44**, 6223–6232 (2003)
33. Yang, Y.: Classical solutions in the Born-Infeld theory. Proc. R. Soc. Lond. Ser. A **456**, 615–640 (2000)

Chapter 6
Projective Geometry

6.1 Extremal Surfaces in Minkowski 3-space

Projective geometry enters the study of partial differential equations of mixed elliptic–hyperbolic type by a very indirect route, beginning with the following geometric variational problem:

The area functional for a smooth surface Σ in $\mathbf{M}^{2,1}$ having graph $z = f(x, y)$ is given by

$$A = \int \int_{\Sigma} \sqrt{\left|1 - f_x^2 - f_y^2\right|}\, dx dy.$$

The surface Σ is *time-like* when $f_x^2 + f_y^2$ exceeds unity and *space-like* when $f_x^2 + f_y^2$ is exceeded by unity. Introducing Lagrange's notation $p = f_x, q = f_y$, the boundary between the space-like and time-like surfaces is the unit circle centered at the origin of coordinates in the pq-plane.

A necessary condition for Σ to be extremal on Minkowski 3-space $\mathbf{M}^{2,1}$ is that its graph $f(x, y)$ satisfy the minimal surface equation in the form [24]

$$\left(1 - p^2\right) q_y + 2pq p_y + \left(1 - q^2\right) p_x = 0. \tag{6.1}$$

This is a quasilinear partial differential equation which is elliptic for space-like surfaces and hyperbolic for time-like surfaces. We can linearize this equation by the hodograph method of Sect. 5.3, obtaining the linear equation

$$\left(1 - p^2\right) \varphi_{pp} - 2pq \varphi_{pq} + \left(1 - q^2\right) \varphi_{qq} = 0. \tag{6.2}$$

(Precisely, apply (5.48), taking $V = px + qy - \varphi(p, q)$, $x = \varphi_p$, $y = \varphi_q$.)

Following Gu [25], we adopt homogeneous coordinates (u, v, w) for $w \neq 0$ and eventually obtain the equation

$$\left[\left(1 - p^2\right) \psi_p\right]_p - 2pq \psi_{pq} + \left[\left(1 - q^2\right) \psi_q\right]_q = 0, \tag{6.3}$$

T.H. Otway, *The Dirichlet Problem for Elliptic-Hyperbolic Equations of Keldysh Type*, Lecture Notes in Mathematics 2043, DOI 10.1007/978-3-642-24415-5_6, © Springer-Verlag Berlin Heidelberg 2012

where $p = -u/w$ and $q = -v/w$. Multiplying both sides of (6.3) by the number $1-p^2-q^2$, we can interpret the resulting equation as the Laplace–Beltrami equation

$$\mathscr{L}_g u = 0,$$

where $\mathscr{L}$ is given by (2.6) and g is the metric on the *extended projective disc*.

6.1.1 The Extended Projective Disc

Beltrami introduced the projective disc $\mathbf{P}^2$ in 1868 as one of the earliest Euclidean models for non-Euclidean space [5]; see also [6, 66]. The geometric idea is to map the hyperbolic plane into the Euclidean plane in such a way that straight hyperbolic lines become straight Euclidean lines (Fig. 6.1). The projective-disc metric amounts to equipping the unit disc centered at the origin of coordinates in $\mathbf{R}^2$ with the distance function

$$ds^2 = \frac{\left(1 - y^2\right) dx^2 + 2xy\,dx\,dy + \left(1 - x^2\right) dy^2}{\left(1 - x^2 - y^2\right)^2}. \tag{6.4}$$

Integrating ds along geodesic lines in polar coordinates, we find that the distance from any point in the interior of the unit disc to the boundary of the disc is infinite, so the unit circle becomes the *absolute*: the curve at projective infinity; c.f. [34], Sect. 9.1. I am reminded of the description by Giambattista della Porta, in 1589, of the *camera obscura*, beginning, "In a small circle of paper, you shall see as it were an epitome of the whole world."

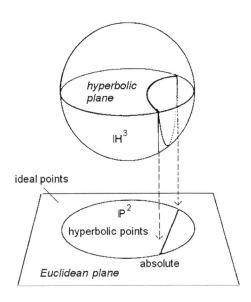

Fig. 6.1 The hyperbolic disc is projected onto the Euclidean disc in such a way that *straight hyperbolic lines* are carried into *straight lines* in $\mathbf{R}^2$. Adapted from Fig. 4.24 of [65]

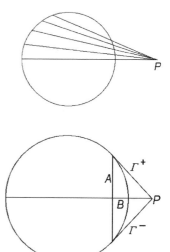

Fig. 6.2 The family of *translated lines* within the unit disc is most conveniently expressed as a family of *rotated lines* which intersect at a point *P* outside the disc; c.f. [65]

Fig. 6.3 Chord *B* is orthogonal to chord *A* if *B* intersects the *polar lines* $\Gamma^{\pm}$ at their intersection, which is the ideal point *P*; c.f. [65]

If we allow the Beltrami metric (6.4) to extend outside the unit disc, then we obtain space-like as well as time-like curves, as the unit disc is, in a certain sense, the projection of a "light cone" onto the xy-plane. Moreover, classical geometric operations within the projective disc are often most conveniently expressed using points that lie beyond the absolute.

For example, certain families of translated lines inside the disc attain their simplest representation as a rotation about an *ideal* point – that is, a point that lies outside the unit disc: Fig. 6.2. (Because points lying inside the unit disc are projective images of $\mathbf{H}^3$, such points are called *hyperbolic*.) In addition, the orthogonality of two chords within the projective disc is most conveniently defined in terms of the intersection of one of the chords with the intersection of the so-called *polar* lines (*i.e.*, tangent lines) of the other. These intersections occur at an ideal point (Fig. 6.3). So the extended projected disc is a natural and mathematically useful generalization of the conventional projective disc.

Recall that the *signature* of a nondegenerate metric is the number of positive and negative eigenvalues. The projective disc is a simple example of a metric which changes from *Riemannian signature* (all eigenvalues positive) to *Lorentzian signature* (one negative eigenvalue and the rest positive) along a smooth curve (the unit circle). Other examples are given in Sect. 6.4.5.

6.2   A Closed Dirichlet Problem Which Is Classically Ill-Posed

A typical domain can be constructed on which the closed Dirichlet problem for an equation of the form (6.3) is ill-posed.

Expressed in polar coordinates, (6.3) assumes the form

$$\left(1 - r^2\right) \phi_{rr} + \frac{1}{r^2} \phi_{\theta\theta} + \left(\frac{1}{r} - 2r\right) \phi_r = 0, \tag{6.5}$$

for $\phi = \phi\left(r, \theta\right)$, provided $r \neq 0$. We define (6.5) over a sector

$$\Omega_s = \{(r, \theta) \,|\, 0 < r_0 \leq r \leq r_1, \, \theta_1 \leq \theta \leq \theta_2\},$$

where $\theta_2 - \theta_1 < \pi$ and the interior of Ω_s includes a segment of the line $r = 1$. For convenience we will take θ_1 to be negative and θ_2 to be positive.

Writing (6.5) in the equivalent form

$$r^2 \left(1 - r^2\right) \phi_{rr} + \phi_{\theta\theta} + r \left(1 - 2r^2\right) \phi_r = 0, \tag{6.6}$$

we show that the Dirichlet problem for this equation on a typical domain has a unique solution if data are given on only the non-characteristic portion of the boundary. This will imply, as in Corollary 3.1, that the Dirichlet problem is over-determined if data are prescribed on the entire boundary.

Equation (6.6) has the same form as (3.21) in the cartesian $r\theta$-plane, if the abscissa is taken to be the line $r = 1$. Thus we could try to apply Theorem 3.1 and its corollary directly to (6.6) (although conditions (3.7) and (3.10) are not satisfied). However, there is some interest in constructing a typical domain for this result in the context of the extended projective disc. So we will initially ignore Theorem 3.1 in approaching the current problem. Eventually, it will become clear how to apply the argument used to prove Theorem 3.1 in the present case.

Define the set $\Omega^+ \subset \Omega_s$, where

$$\Omega^+ = \left\{(r, \theta) \in \mathbf{R}^2 \,|\, r_0 < \varepsilon \leq r < 1, -\theta_0 \leq \theta \leq \theta_0\right\}.$$

We will choose the domain Ω of (6.6) to be the region enclosed by the annular sector Ω^+ and the intersecting lines tangent to the points $(r, \pm\theta) = (1, \pm\theta_0)$; see Fig. 6.4.

Let Ω_0 be the triangular region bounded by the vertical chord γ_0 given in cartesian coordinates by

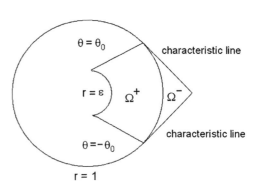

Fig. 6.4 A domain on which the closed Dirichlet problem is classically ill-posed

$$\gamma_0 = \{(x, y) \in \mathbf{R}^2 | x = x_0, x_0 > 0, y^2 \le 1 - x_0^2\}$$

and the two polar lines of γ_0. (Recall that the *polar* lines of a chord are the tangent lines to the unit circle at its two points of intersection with the chord.) Then $\Omega = \Omega^+ \cup \Omega_0$.

In the following we denote by Ω^- the subdomain of Ω_0 consisting of ideal points, and by ν the arc of the unit circle lying between the points $(r, \pm \theta) = (1, \pm \theta_0)$. Because ε is a fixed number greater zero, the mapping from the region Ω in the polar $r\theta$-plane (the cartesian xy-plane) to its image in the cartesian $r\theta$-plane is well-defined, and we shall call the image of this mapping Ω as well.

Theorem 6.1 ([60]). *Considering (6.6) as an equation on the subdomain Ω of the cartesian $r\theta$-plane, any sufficiently smooth solution ϕ of (6.6) taking values $f(r, \theta) \in C^2(\partial\Omega)$ on the boundary segment $\partial\Omega \backslash \Omega^-$ is unique.*

The proof of Theorem 6.1 is exactly analogous to the proof of Theorem 3.1. Suppose that there are two solutions satisfying the boundary conditions. Subtraction yields a solution, which we also denote by ϕ, satisfying homogeneous boundary conditions. Define the functional

$$I = \int^{(r,\theta)} \psi_1 d\theta + \psi_2 dr, \tag{6.7}$$

where

$$\psi_1 = r^2 \left(1 - r^2\right) \phi_r^2 - \phi_\theta^2 \tag{6.8}$$

and

$$\psi_2 = -2\phi_r \phi_\theta, \tag{6.9}$$

on the cartesian $r\theta$-plane. A solution of (6.6) will be considered to be sufficiently smooth in the sense of Theorem 6.1 if either of the smoothness conditions in the Remark following Theorem 3.1 is satisfied. We compute

$$\psi_{2\theta} - \psi_{1r} = -2\phi_r \left[r^2 \left(1 - r^2\right) \phi_{rr} + \phi_{\theta\theta} + r \left(1 - 2r^2\right) \phi_r\right] = 0, \tag{6.10}$$

and observe that there is a function $\chi(r, \theta)$ such that $\chi_\theta = \psi_1$ and $\chi_r = \psi_2$. Now we proceed, as in the proof of Theorem 3.1, to show that the difference of the two solutions must be identically zero on the domain, by applying the maximum principle in the elliptic region and by integrating along characteristic lines in the hyperbolic region; see [60], Sect. 3.1, for details.

6.2.1   Will These Methods Work for Magnanini and Talenti's Model?

Yes. Consider (5.60) on a domain Ω, considered as the union $\Omega = \Omega_+ \cup \Omega_-$, where

$$\Omega_+ = \{(r, \theta) | 1 \le r < R, -\pi < \theta_0 \le \theta \le \theta_1 < \pi\};$$

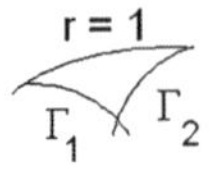

Fig. 6.5 The domain is bounded in the hyperbolic region by characteristics Γ_1 and Γ_2, which are intersecting arcs of the circles defined by (6.11)

Ω_- is any subdomain of the annulus $0 < \varepsilon_1 \leq r < 1$ the upper boundary of which is the unit circle and the intersection of two characteristic lines with the points $(1, \theta_0)$, $(1, \theta_1)$ and with each other. In the following we prove that it is always possible to find such an intersection of characteristic lines, because the characteristic lines form a family of circles of radius 1/2, which are all tangent to the interior of the unit circle; c.f. Fig. 6.5.

Proposition 6.1. *[c.f. [50]] The circle $r = 1$ and the origin of coordinates are both envelopes of the family of characteristic lines for (5.60).*

Proof. The characteristic lines of the cartesian form of equation (5.60),

$$\left[\left(p^2 + q^2\right)^2 - p^2\right] u_{pp} - 2pq u_{pq} + \left[\left(p^2 + q^2\right)^2 - q^2\right] u_{qq} = 0,$$

have the form $F(p, q; \Theta) = 0$ for

$$F(p, q; \Theta) = p^2 + q^2 - p \cos \Theta - q \sin \Theta, \qquad (6.11)$$

where Θ can be treated as a real parameter ([50], Sect. 5). Let $\mathscr{F}$ be any 1-parameter family of smooth planar curves defined by the equation $F(p, q; \Theta) = 0$. Then the envelope C_F of the family $\mathscr{F}$ belongs to a subset of the set of points (p, q) satisfying $F = 0$ and the equation $F_\Theta = 0$, provided C_F is sufficiently smooth. Applying this criterion with F given by the characteristic family (6.11), we find that the equations $F = 0$ and $F_\Theta = 0$ are satisfied if p and q both vanish identically, and also if $p = \cos \Theta$ and $q = \sin \Theta$. So the unit circle centered at the origin of the pq-plane and the origin itself are both envelopes. This completes the proof.   □

We can express (5.60) *itself* as an equation in cartesian coordinates if we consider Ω to be a domain of the euclidean half-plane $r \geq \varepsilon_1 > 0$. In this case the line $r = 1$ remains an envelope of the characteristic equations to (5.60).

Theorem 6.2. *Consider (5.60) on the domain Ω of the cartesian $r\theta$-plane. Any twice-differentiable solution taking twice-differentiable values on the boundary segment $\partial(\Omega/\Omega_-)$ is unique.*

Proof. Proceed exactly as in the proof of Theorem 3.1, applying a maximum principle in the elliptic region and integrating along characteristic lines in the hyperbolic region (Fig. 6.5).   □

6.3 An Open Dirichlet Problem Which Is Weakly Well-Posed

In this section we show the weak existence of solutions to a large class of Guderley–Morawetz problems. Recall that in such problems, the solution is prescribed on the non-characteristic boundary of the domain. We show that weak solutions exist for the inhomogeneous equation under very mild hypotheses on the forcing function. We will prove our results for a natural generalization of (6.3).

6.3.1 The Equations

Equation (2.6) on the Beltrami metric (6.4) of extended $\mathbf{P}^2$ takes the form

$$\left(1 - x^2 - y^2\right) \left\{ \left[\left(1 - x^2\right) u_1\right]_x - xy \left(u_{1y} + u_{2x}\right) + \left[\left(1 - y^2\right) u_2\right]_y \right\} = 0. \quad (6.12)$$

If in addition, we have the condition

$$u_{1y} - u_{2x} = 0, \quad (6.13)$$

then there exists a scalar potential function $\psi\,(x, y)$ such that, locally, $u_1 = \psi_x$ and $u_2 = \psi_y$. The scalar function ψ then satisfies the second-order scalar equation (6.3), which we write in the form

$$\left[\left(1 - x^2\right) \psi_x\right]_x - 2xy\,\psi_{xy} + \left[\left(1 - y^2\right) \psi_y\right]_y = 0 \quad (6.14)$$

on the complement of the absolute in extended $\mathbf{P}^2$.

Replacing (6.13) by condition

$$u_{1y} - u_{2x} - \left(u_1 \Gamma_2 - u_2 \Gamma_1\right) = 0, \quad (6.15)$$

where Γ_1 and Γ_2 are prescribed functions of x and y, we obtain from (6.12) a new system, consisting of (6.15) and the equation

$$\frac{1}{\sqrt{|g|}} \frac{\partial}{\partial x^i} \left(g^{ij} \sqrt{|g|} u_j\right) = 0, \quad (6.16)$$

where i and j range from 1 to 2 and, as in (2.6), repeated indices are summed from 1 to 2. The system (6.16), (6.15), is equivalent to the Hodge–Frobenius equations (5.76), (5.77) for 1-forms on $\mathbf{P}^2$; here g_{ij} is the metric tensor on the extended projective disc. (Precisely, it is equivalent to the form of these equations which was introduced in [55], in which the right-hand side of (5.76) is set to zero. See also [57].) Recall that in substituting condition (6.15) for condition (6.13) we are in

effect substituting a completely integrable field for a conservative one; c.f. Tables
5.1 and 5.2. Complete integrability is not a sufficiently strong condition to imply
the existence of a scalar potential but is, in effect, the mildest weakening of the
hypothesis that such a potential exists (Sect. 5.6.1).

We obtain the coupled system consisting of the equation

$$\left[\left(1 - x^2\right) u_1\right]_x - xy \left(u_{1y} + u_{2x}\right) + \left[\left(1 - y^2\right) u_2\right]_y = 0 \qquad (6.17)$$

and the additional constraint (6.15). In fact, applying condition (6.15) to the term
u_{2x} in (6.17), we arrive at the slightly simpler system consisting of the equations
(2.13) with $\mathbf{f} = 0$,

$$(L\mathbf{u})_1 = \left[\left(1 - x^2\right) u_1\right]_x - xy \left[2u_{1y} - \left(u_1 \Gamma_2 - u_2 \Gamma_1\right)\right] + \left[\left(1 - y^2\right) u_2\right]_y, \quad (6.18)$$

and

$$(L\mathbf{u})_2 = \left(1 - y^2\right) \left[\left(u_{1y} - u_{2x}\right) + \Gamma_1 u_2 - \Gamma_2 u_1\right]. \qquad (6.19)$$

Provided $y^2 \neq 1$, this alternative differs from the case $\Gamma_1 = \Gamma_2 = 0$ only in its
lower-order terms.

Now we will introduce a significant modification. Equations (2.13), (6.18), and
(6.19) comprise a system of the form

$$F_1 + (\text{lower-order})_1 = 0, \qquad (6.20)$$

$$F_2 + (\text{lower-order})_2 = 0, \qquad (6.21)$$

where F_1 and F_2 denote the higher-order terms of the equations. This system implies
the single equation

$$F_1 + (\text{lower-order})_1 = F_2 + (\text{lower-order})_2, \qquad (6.22)$$

but of course (6.22) does not imply the system (2.13), (6.20), and (6.21). So proving
the existence of solutions to a boundary value problem for (6.22) would not be
equivalent to proving the analogous result for (2.13), (6.20), and (6.21).

The advantage to considering (6.22) instead of the system (6.20), (6.21) is that it
allows us to move the lower-order terms around for our convenience. For example,
(6.22) is equivalent to the equation

$$F_1 - (\text{lower-order})_2 = F_2 - (\text{lower-order}))_1.$$

This allows us to prove an assertion of the form

$$F_1 - (\text{lower-order})_2 = f_1 \qquad (6.23)$$

$$F_2 - (\text{lower-order})_1 = f_2 \qquad (6.24)$$

where f_1, f_2 are a class of sufficiently smooth prescribed functions which presumably includes the case which makes (2.13), (6.23), and (6.24) equivalent to (2.13), (6.20), and (6.21). (Note that they are *not* equivalent if $f_1 = f_2 = 0$.)

Thus we will restrict our attention to the system (2.13), with

$$(L\mathbf{u})_1 = \left[\left(1 - x^2\right) u_1\right]_x - 2xyu_{1y} + \left[\left(1 - y^2\right) u_2\right]_y + \gamma_2 u_1, \qquad (6.25)$$

$$(L\mathbf{u})_2 = \left(1 - y^2\right) \left(u_{1y} - u_{2x}\right) + \gamma_1 u_2, \qquad (6.26)$$

where we are particularly interested in the choice

$$\gamma_i = \left(xy + 1 - y^2\right) \Gamma_i, \ i = 1, 2.$$

In accordance with the preceding discussion, the terms γ_i must be sufficiently smooth. In fact, we will require them to be bounded, integrable, nonvanishing, and to satisfy a sign condition (Theorem 6.3).

Characteristic lines for this system satisfy the differential equation

$$\left(1 - y^2\right) dx^2 + 2xy\,dx\,dy + (1 - x^2)dy^2 = 0. \qquad (6.27)$$

Equation (6.27) has solutions

$$x \cos \theta + y \sin \theta = 1, \qquad (6.28)$$

where, as is conventional, we take θ to be the angle between the radial vector and the positive x-axis. Solutions of (6.27) correspond geometrically to the family of tangent lines to the unit circle centered at the origin of $\mathbf{R}^2$.

6.3.2   The Associated Spaces

Denote by Ω a region of the plane for which part of the boundary $\partial\Omega$ consists of a family $\tilde{\mathcal{G}}$ of curves composed of points satisfying (6.27) and the remainder $C = \partial\Omega \backslash \tilde{\mathcal{G}}$ of the boundary consists of points (x, y) which do not satisfy (6.27). In addition to the requirements that Ω contain an arc of the unit circle and that y^2 is bounded above away from unity, we require that any points satisfying the equation

$$xy + 1 - y^2 = 0 \qquad (6.29)$$

lie in the complement of Ω. Let $\partial\Omega$ be oriented in the counter-clockwise direction and have piecewise continuous tangent.

Let the parameter θ of (6.28) lie in the interval $[0, \pi/4]$ and let Ω occupy the region of the first and fourth quadrants bounded by the characteristic line

Fig. 6.6 The domain in the case in which the curve C is a chord of the unit circle

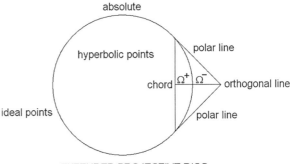

EXTENDED PROJECTIVE DISC

$$\tilde{\mathcal{G}}_1 : x \cos \theta + y \sin \theta = 1,$$

the characteristic line

$$\tilde{\mathcal{G}}_2 : x \cos \theta - y \sin \theta = 1,$$

and a smooth curve C. Let C intersect the lines $\tilde{\mathcal{G}}_1$, $\tilde{\mathcal{G}}_2$ at two distinct points c_1, c_2, respectively. Assume that $dy \leq 0$ on C if γ_1 and γ_2 are negative definite and that $dy \geq 0$ if γ_1 and γ_2 are positive definite. (This assumption generally determines the choice of quadrant in which Ω lies.) In particular, the hypotheses are satisfied if C is a chord of the projective disc for which $\tilde{\mathcal{G}}_1$ and $\tilde{\mathcal{G}}_\epsilon$ are the corresponding polar lines: Fig. 6.6. (Figure 6.6 implies that the Laplace–Beltrami equation is elliptic only on *hyperbolic* points! While this is superficially confusing, it is clear that the word "hyperbolic" in this context refers to $\mathbf{H}^3$.)

Note that if θ is close to zero, then y is close to 0. In that case it is easy to see that condition (6.29) is not satisfied in Ω, as required; c.f. [56].

We seek solutions of (2.13), (6.25), and (6.26) which satisfy the boundary condition

$$u_1 \frac{dx}{ds} + u_2 \frac{dy}{ds} = 0, \qquad (6.30)$$

where s denotes arc length, on the non-characteristic part C of the domain boundary. Because the tangent vector $\mathbf{T}$ on C is given by

$$\mathbf{T} = \frac{dx}{ds}\mathbf{i} + \frac{dy}{dx}\mathbf{j},$$

a geometric interpretation of this boundary condition is that the dot product of the vector $\mathbf{u} = (u_1, u_2)$ and the tangent vector to C vanishes, *i.e.*, $\mathbf{u}$ is normal to the boundary $\partial \Omega$ on the boundary section C. Thus condition (6.30) prescribes homogeneous open Dirichlet conditions.

Define U to be the vector space consisting of all pairs of measurable functions $\mathbf{u} = (u_1, u_2)$ for which the weighted L^2 norm

$$\|\mathbf{u}\|_* = \left[\int\!\!\int_\Omega \left(|\gamma_1(x,y)|\, u_1^2 + |\gamma_2(x,y)|\, u_2^2 \right) dx\, dy \right]^{1/2}$$

is finite. Denote by W the linear space defined by pairs of functions $\mathbf{w} = (w_1, w_2)$ having continuous derivatives and satisfying:

$$w_1\, dx + w_2\, dy = 0 \tag{6.31}$$

on $\tilde{\mathscr{G}} = \tilde{\mathscr{G}}_1 \cup \tilde{\mathscr{G}}_2$;

$$w_1 = 0 \tag{6.32}$$

on C; and

$$\int\!\!\int_\Omega \left[|\gamma_1(x,y)|^{-1} \left(L^*\mathbf{w} \right)_1^2 + |\gamma_2(x,y)|^{-1} \left(L^*\mathbf{w} \right)_2^2 \right] dx\, dy < \infty.$$

Here

$$\left(L^*\mathbf{w} \right)_1 = \left[(1 - x^2)\cdot w_1 \right]_x - 2xy w_{1y} + \left[(1 - y^2) w_2 \right]_y - \gamma_2(x,y)\, w_1,$$

and

$$\left(L^*\mathbf{w} \right)_2 = (1 - y^2)\left(w_{1y} - w_{2x} \right) - \gamma_1(x,y)\, w_2.$$

Define the Hilbert space H to consist of pairs of measurable functions $\mathbf{h} = (h_1, h_2)$ for which the norm

$$\|\mathbf{h}\|^* = \left[\int\!\!\int_\Omega \left(|\gamma_1(x,y)|^{-1} h_1^2 + |\gamma_2(x,y)|^{-1} h_2^2 \right) dx\, dy \right]^{1/2}$$

is finite.

We say that $\mathbf{u}$ is a *weak solution* of the system (2.13), (6.25), (6.26), (6.30) on Ω if $\mathbf{u} \in U$ and for every $\mathbf{w} \in W$,

$$-(\mathbf{w}, \mathbf{f}) = \left(L^*\mathbf{w}, \mathbf{u} \right),$$

where

$$(\mathbf{w}, \mathbf{f}) = \int\!\!\int_\Omega (w_1 f_1 + w_2 f_2)\, dx\, dy.$$

6.3.3 Weak Existence

Theorem 6.3. *Let the domain Ω satisfy the hypotheses of Sect. 6.3.2. Suppose that the bounded functions γ_1 and γ_2 are definite with the same sign. Then there exists a* *weak solution of the boundary value problem (2.13), (6.25), (6.26), (6.30) on Ω for every* **f** $\in$ *H.*

Remark. The limiting case $\gamma_1 = \gamma_2 \equiv 0$ was treated in Sect. II, Case 3, of [58, 59]. If γ_1 and γ_2 are both positive and bounded below away from zero, or both negative and bounded above away from zero, then H can be replaced by L^2.

Proof. Following [54], we derive the customary fundamental inequality, that there is a $k \in \mathbf{R}^+$ such that $\forall \mathbf{w} \in W$,

$$k \|\mathbf{w}\|_* \leq \|L^* \mathbf{w}\|^* . \qquad (6.33)$$

We derive this inequality by choosing a scalar multiplier a, computing the L^2 inner product $(L^* \mathbf{w}, a\mathbf{w})$, and integrating by parts. Choose a to be 1 if γ_1 and γ_2 are negative definite; choose a to be -1 otherwise. Applying Green's Theorem to the derivatives of products as in Proposition 2.3, we obtain an identity of the form

$$\left(L^* \mathbf{w}, a\mathbf{w}\right) = \mathscr{I} + a \int_\Omega \left(|\gamma_1| w_1^2 + |\gamma_2| w_2^2\right) dx dy,$$

where

$$\mathscr{I} = \int_{\partial\Omega} \frac{a}{2} \left[(1 - x^2) w_1^2 dy + 2xy w_1^2 dx\right]$$
$$- \int_{\partial\Omega} a \left[(1 - y^2) w_1 w_2 dx + \frac{1}{2}(1 - y^2) w_2^2 dy\right].$$

Because w_1 vanishes identically on C by (6.32), the boundary integral is nonnegative on C by the hypothesis on the signs of a and $dy_{|C}$. On the characteristic curves $\tilde{\mathscr{G}}$, (6.31) implies that [56]

$$\mathscr{I}_{|\tilde{\mathscr{G}}} = \int_{\tilde{\mathscr{G}}} \frac{a}{2} \left\{(1 - x^2) w_1^2 dy + [2xy w_1^2 - (1 - y^2) w_1 w_2] dx\right\} .$$

In fact,

$$\mathscr{I}_{|\tilde{\mathscr{G}}} = \int_{\tilde{\mathscr{G}}} \frac{a}{2} \left[(1 - x^2) w_1^2 \left(\frac{dy}{dx}\right) + 2xy w_1^2 - (1 - y^2) w_1 w_2\right] dx$$
$$= \int_{\tilde{\mathscr{G}}} \frac{a}{2} \left[-(1 - x^2) w_1 w_2 \left(\frac{dy}{dx}\right)^2 + 2xy w_1^2 - (1 - y^2) w_1 w_2\right] dx.$$

Equation (6.27) can be written in the form

$$-(1 - x^2) \left(\frac{dy}{dx}\right)^2 = 2xy \frac{dy}{dx} + 1 - y^2,$$

implying

$$\mathcal{I} = \int_{\mathcal{G}} \frac{a}{2} \left[2xy \frac{dy}{dx} + 1 - y^2 \right] w_1 w_2 dx$$
$$+ \int_{\mathcal{G}} \frac{a}{2} \left[2xyw_1(-w_2 \frac{dy}{dx}) - (1 - y^2)w_1 w_2 \right] dx = 0.$$

This establishes the fundamental inequality (6.33).

The fundamental inequality allows us to apply the Riesz Representation Theorem as in the preceding chapters. We obtain an element $\mathbf{h} \in H$ for which

$$- (\mathbf{w}, \mathbf{f}) = -(L^* \mathbf{w}, \mathbf{h})^*, \tag{6.34}$$

where the product on the right is the inner product on H. Defining $\mathbf{u} = (u_1, u_2)$, where

$$u_1 = \frac{h_1}{|\gamma_1|}, \; u_2 = \frac{h_2}{|\gamma_2|},$$

we obtain

$$- (L^* \mathbf{w}, \mathbf{h})^* = (L^* \mathbf{w}, \mathbf{u}) \tag{6.35}$$

for $\mathbf{u} \in U$. Substituting (6.35) into (6.34) completes the proof. □

6.4 Remarks on Some Related Topics

We mention several topics related to the subject of this chapter which we have not included, as they require their own array of technical methods. For example, the work by Hua and his students (Sect. 6.4.1) requires techniques of harmonic analysis; the work by Gu (Sect. 6.4.2) requires delicate extensions of the theory of symmetric positive operators. Other applications would require an extensive discussion of cosmology in order to treat them in detail (Sect. 6.4.5).

6.4.1 Tricomi Problems

L-K. Hua has identified three classes of *Tricomi problem* (Sect. 2.3) which have interest for the Laplace–Beltrami equation on extended $\mathbf{P}^2$. If values of the solution are prescribed on characteristic lines and on a curve C in the elliptic region of the equation, then either:

1. C intersects the characteristic lines
2. C does not intersect the characteristic lines
3. C is tangent to the characteristic lines

Fig. 6.7 The case of non-
intersecting characteristics;
see Sect. 4 of [36]

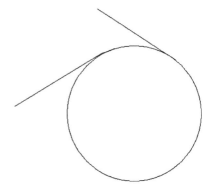

By contrast, the open Dirichlet problem considered in Sect. 6.3 is a Guderley–
Morawetz problem, in which data are not given on characteristic lines but rather on
the elliptic and non-characteristic hyperbolic boundaries.

In each of the Problems 1–3 one may vary the geometry of the characteristic
boundary: The characteristics may intersect at at a zero angle, to form a single
tangent line (c.f. [56]); they may intersect at an acute angle, as in Sect. 6.3 (see
Figs. 6.3 and 6.6); or they may not intersect at all, as in Fig. 6.7.

Using techniques of harmonic analysis and classical function theory, Hua solved
a Tricomi problem of the form 1 [36], and X-H. Ji and D-Q. Chen solved Tricomi
problems of the form 2 and 3 [38].

6.4.2   Equations in n Independent Variables

All the elliptic–hyperbolic equations considered in this review are in two independent variables. Even in the more accessible case of equations of Tricomi type, the
theory for n-dimensional equations, where n exceeds two, is very sparse.

However, C-H. Gu has considered boundary value problems for a large class
of elliptic–hyperbolic equations in n variables [27]. This class includes equations
having the form

$$L\varphi \equiv \left(\delta^{ij} - x^i x^j \right) \partial_{ij}\varphi + 2ax^i \partial_i \varphi - a\,(a+1)\,\varphi = f, \qquad (6.36)$$

where $i, j = 1, 2, \ldots, n$, on a bounded domain Ω for $\mathbf{R}^n$. Obviously, (6.36)
degenerates, in the case $a = -1$ and $n = 2$, to the Laplace–Beltrami equation on
extended $\mathbf{P}^2$. Unfortunately, this choice of a is explicitly excluded by Gu's existence
theorems. However, applications of (6.36) to higher-dimensional wave equations,
including certain cases in which explicit solutions can be written down, are given at
the end of Sect. 6.4.4.

Gu imposes a hypothesis on the boundary which is quite different from those
imposed in this review, and which would exclude the analogies to extended $\mathbf{P}^2$ which

we have emphasized: In addition to being smooth, it is assumed that the tangent planes of $\partial\Omega$ do not intersect the unit sphere.

Theorem 6.4 (Gu [26, 27]). *For $s > 1$, let $f \in H^{1,s}(\Omega)$ and $a > -(n/2) + s$. Then there exists a unique solution $\varphi \in H^{1,s+1}(\Omega)$ to (6.36) which satisfies the boundary conditions*

$$\varphi_{|\partial\Omega} = 0, \quad \frac{\partial\varphi}{\partial n}\bigg|_{\partial\Omega} = 0, \tag{6.37}$$

where $\partial/\partial n$ indicates differentiation in the direction of the outward-pointing normal. In the special case $s \geq (n/2) + 2$, there is a classical solution to (6.36), (6.37).

In addition, under the conditions of the theorem the solution to (6.36) is unique whenever $a < -n/2$.

6.4.3   Very Brief Historical Remarks

Equation (6.36) is a special case of a more general class of equations considered by Gu, which satisfy [27]

$$h^{ij}\partial_{ij}\varphi + p^i\partial_i\varphi + q\varphi = f, \tag{6.38}$$

for $i, j = 1, \ldots, n$, where h^{ij} and p^i are given, sufficiently smooth functions. We will only consider one aspect of the analysis of (6.38), which is of historical interest.

Writing, by a partition of unity,

$$a^{ij} = h^{ij} + \sigma^i\sigma^j,$$

(6.38) is transformed into the form

$$\left(a^{ij} - \sigma^i\sigma^j\right)\partial_{ij}\varphi + p^i\partial_i\varphi + q\varphi = f, \tag{6.39}$$

where a^{ij} can be chosen to be a positive-definite matrix. This equation is elliptic under the strict inequality

$$a_{ij}\sigma^i\sigma^j < 1,$$

where a_{ij} is the inverse of the matrix a^{ij}, and hyperbolic under the reverse strict inequality. Introducing the coordinate transformation

$$u_0 = \lambda\varphi - \sigma^i\partial_i\varphi,$$

$$u_i = \partial_i\varphi,$$

where λ is an arbitrary nonvanishing function, conditions can be imposed under which (6.39) is equivalent to a first-order symmetric positive system.

In deriving sufficient conditions for the existence of classical solutions to boundary value problems for (6.39), Gu imposes the additional hypotheses that the region $\overline{\Omega}$ is simply connected and the trajectories of the vector field σ^i cover the whole of $\overline{\Omega}$. This is essentially equivalent to the hypothesis that the domain is star-shaped with respect to the vector field σ^i, c.f. Sect. 3.5.1. Gu's paper [27], published in 1981, was an early application of that hypothesis to the theory of elliptic–hyperbolic equations.

Starlike boundaries in the sense of Sect. 3.5.1 can be found in even earlier literature. For example, in Katsanis' doctoral dissertation (c.f. [39, 40]) various boundary conditions of the form (3.45) are applied. As is illustrated by Theorem 3.4, such conditions are natural in the context of Friedrichs' theory of symmetric positive operators, which is also the context in which they are applied in [27]. But Katsanis does not connect these conditions with any geometric interpretation or even with the existence of a vector field, whereas Gu explicitly interprets his condition in terms of the flow of a vector field. The association of inequalities of the form (3.45) with vector fields can be found in Sect. 2 of Didenko's paper [13], published in 1973. But the full power of these ideas would not be apparent until the recent work of Lupo and Payne [49] and Lupo, Morawetz, and Payne [47, 48].

6.4.4 The Busemann Equation

The original context for (6.3) was fluid dynamics. (Moreover, its characteristic (6.27) seems to have been introduced by the mathematical physicist G. G. Stokes in the same Smith's Prize examination in which he introduced Stokes' Theorem; see [67]). Busemann [8] noticed that the substitution

$$\theta = \arctan \frac{q}{p}, \ r = \sqrt{p^2 + q^2}$$

transforms (6.3) into a wave equation having the form

$$\psi_{\lambda\mu} = 0$$

for $r > 1$ and

$$\lambda = \theta + \arccos\left(\frac{1}{r}\right), \ \mu = \theta - \arccos\left(\frac{1}{r}\right);$$

see also Sect. 3 of [25].

Conversely, the wave equation in the form

$$u_{tt} = c_0^2 \left(u_{xx} + u_{yy}\right)$$

can be transformed into (6.3) by the substitution

$$p = \frac{x}{c_0 t}, \quad q = \frac{y}{c_0 t};$$

see, e.g., [37,41].

The term *Busemann's equation* is often applied to (6.3) in the form (6.2); but it can be applied more generally to all the equations studied up until now in this chapter. A generalized form of Busemann's equation

$$\left(\delta^{ij} - x^i x^j\right) \varphi_{x^i x^j} + 2a x^i \varphi_{x_i} - b\varphi = f, \tag{6.40}$$

where $i, j = 1, \ldots, n$ and $x = \left(x^1, \ldots, x^n\right) \subset \Omega \subset \mathbf{R}^n$, was studied by Gu [27], who proved the existence of classical solutions to certain homogeneous boundary value problems on domains having smooth boundary. The equation has been recently been studied by Xu, Xu, and Chen [74], who find conditions under which (6.40) has H^1 solutions under boundary conditions given on a nonsmooth boundary. Eigenvalue problems for (6.40) are also addressed in [74].

As the preceding discussion suggests, solutions to the Busemann-type equation (6.36) for $f = 0$ have an interpretation as waves in $\mathbf{R}^{n+1}$; these solutions are homogeneous of degree $a + 1$. If $n = 1$, then Gu's solutions have the explicit form

$$u(x) = \alpha |1 - x|^{a+1} + \beta |1 + x|^{a+1},$$

where α and β are constants. See [23, 26], and Sect. 3 of [27].

6.4.5 Metrics of Mixed Riemannain–Lorentzian Signature

The projective disc is only one example of a metric that changes signature along a smooth curve. Many others occur in connection with general relativity and cosmology. The associated Laplace–Beltrami equations tend to be of Tricomi type. Nevertheless, we include a brief survey of signature change in the recent physics literature, following Sect. 6 of [60].

1. *Special relativity*: Stationary waves on Minkowski space-time, in a reference frame rotating with constant angular velocity ω with respect to another reference frame, can be written in cylindrical coordinates (ρ, φ, z) as the elliptic–hyperbolic equation [63, 64]

$$\frac{1}{\rho} \left(\rho u_\rho\right)_\rho + \left(\frac{1}{\rho^2} - \omega^2\right) u_{\varphi\varphi} + u_{zz} = 0. \tag{6.41}$$

Taking the initial metric to be the standard Minkowski line element written in cylindrical coordinates

$$ds_{old}^2 = -dt^2 + d\rho^2 + \rho^2 d\phi^2 + dz^2,$$

one obtains (6.41) by introducing a reference frame for the observer, which rotates relative to Minkowski space at a constant angular velocity ω. In terms of the azimuthal angle $\varphi = \phi + \omega t$, we obtain a composite line element of the form

$$ds_{new}^2 = -dt^2 + d\rho^2 + \rho^2 (d\varphi - \omega dt)^2 + dz^2.$$

Writing the Laplace–Beltrami equation on the associated metric yields (6.41).

2. *Binary black hole space-times with a helical killing vector* [42]: This model is a generalization of the preceding example; see also [35,68]. Note that boundary value problems for this model tend to have so-called *Sommerfeld* boundary conditions, which are similar to the boundary conditions of Theorem 5.2.

3. *Quantum cosmology*: The best known examples of metrics which change signature arise from the controversial *Hartle–Hawking hypothesis* [28], that the universe might have originated as a manifold having Euclidean signature and subsequently undergone a transition to a model having Lorentzian signature across a hypersurface which was space-like as seen from the Lorentzian side. (In using this terminology we acknowledge that certain metrics which are called *Euclidean* by physicists would be called *Riemannian* by geometers; see footnote 2 of [21].) In two dimensions, the following common variants can be isolated [64]:

i) Continuous change of signature:

$$ds^2 = -t dt^2 + dz^2;$$

ii) Discontinuous change of signature:

$$ds^2 = -z^{-1} dt^2 + dz^2;$$

iii) Continuous change of signature with a curvature singularity:

$$ds^2 = -z dt^2 + dz^2;$$

see also [15].

In four dimensions one also finds [1] a change in the metric from Lorentzian to Kleinian signature across the line $z = 0$:

$$ds^2 = -dt^2 + dx^2 + dy^2 + z dz^2.$$

The distinction between examples *ii)* and *iii)* suggests that certain operators on Riemannian–Lorentzian metrics degenerate rather than blow up at the change of signature. Such metrics have been studied by mathematicians as well as physicists [10]; see also [9] and the discussion in Sect. 5 of [62].

Although many of the criticisms of the Hartle–Hawking model arise from its physical implications, controversy also arises from mathematical ambiguities in geometric analysis on mixed Riemannian–Lorentzian metrics and from the wide range of potential singularities of such metrics. See [61] and the references therein for a recent discussion of the physical predictions, and [14, 15, 18, 19, 30, 32, 33, 45] for discussions of the mathematical ambiguities. It has also been noted [17] that singularities similar to those that are associated with the Hartle–Hawking transition can also arise in classical relativity, as in Example 1, above; see also [11, 12, 16, 31, 43, 44].

4. *Repulsive singularities in four-dimensional extended supergravity* [20]: Mathematically, this model is essentially a combination of 3*ii)* and 3*iii)*. Precisely, define a surface in spherical coordinates having conventional distance element [64]

$$d\Sigma^2 = dr^2 + r^2 \left(d\theta^2 + \sin^2\theta d\phi^2\right).$$

From this we construct the space-time metric

$$ds_{inner}^2 = |e^{-2V}|dt^2 + |e^{2V}|d\Sigma^2, \ 0 < r < r_0,$$

and

$$ds_{outer}^2 = -e^{-2V}dt^2 + e^{2V}d\Sigma^2, \ r_0 < r < \infty,$$

where

$$e^{4V} = 1 + \sum_{n=1}^{4} c_n r^{-n}$$

for $c_1 > 0$. We choose c_2, c_3, and c_4 in such a way that e^{4V} is negative for $r < r_0$ and positive for $r > r_0$.

Inside the *horizon* $r = r_0$, this space-time is "Euclidean" (that is, Riemannian), but outside the horizon it is Lorentzian. The curvature is singular at the horizon itself.

5. *Brane worlds*: A *brane* is a submanifold of a higher-dimensional space-time, called a *bulk*. Branes are generally represented as uniformly time-like, but they need not be time-like everywhere [51] and provide a natural context for signature change. Mixed Euclidean–Lorentzian branes can be constructed in such a way that both the bulk and the brane are regular; but if viewed from within the brane, the change of signature appears as a curvature singularity [21, 52]. This structure can provide a kinematic model for both the apparent big bang singularity and the apparent accelerated expansion of the universe. In such a model, no hypothesis of dark energy is needed to account for accelerated expansion [53].

6. *Acoustic models*: Recently, analogies for the dynamical equations for light in curved space-time have been constructed using equations from models of condensed matter, for example, Bose–Einstein condensates with a sink or a vortex [4, 46, 70–73]. The analogies are drawn between acoustic waves in matter and

wave equations on mixed Riemannian–Lorentzian manifolds. Mathematically, these analogies are somewhat reminiscent of those between gas dynamics and extremal surfaces discussed in Sect. 5.6.1.

Acoustic models are based on the observation that an acoustic wave has a relation to the flow in which it propagates which is partly analogous to the relation between a light wave and the ambient space-time. One obtains from this analogy a kinematic model for relativistic effects, in that if the flow becomes supersonic, an acoustic wave emitted downstream from a listener will be trapped in an analogous way to the trapping of light inside a black hole with respect to an external observer. Such analogies can be traced back at least to the flow metrics of [7], and possibly even to the electrodynamics of [22]. See [3] for a review.

Signature change has also been framed in the context of spinor cosmology; see, e.g., [69].

An elliptic–hyperbolic system of equations associated with the Einstein equations is studied in [2]. But in that example an elliptic gauge-fixing condition is coupled to the hyperbolic evolution equations. The metric is Lorentzian, so the signature is fixed and the resulting system is qualitatively different from the examples in this section. Nor is the analysis similar to the methods discussed in this text. Similar remarks apply to the Davey–Stewartson system of hydrodynamics in which an elliptic equation is coupled to a hyperbolic equation; see, e.g., [29].

References

1. Alty, L.J.: Kleinian signature change. Class. Quantum Grav. **11**, 2523–2536 (1994)
2. Andersson, L., Moncrief, V.: Elliptic-hyperbolic systems and the Einstein equations. Ann. Henri Poincaré **4**, 1–34 (2003)
3. Barceló, C., Liberati, S., Visser, M.: Analogue gravity. Living Rev. Rel. **8**, 12 (2005)
4. Barceló, C., Liberati, S., Sonego, S., Visser, M.: Causal structure of acoustic spacetimes. New J. Phys. 186 (2004)
5. Beltrami, E.: Saggio di interpretazione della geometria non-euclidea. Giornale di Matematiche **6**, 284–312 (1868)
6. Beltrami, E.: Teoria fondamentale degli spazii di curvatura costante. Annali Mat. Pura Appl. ser. 2. **2**, 232–255 (1868)
7. Bers, L.: Mathematical Aspects of Subsonic and Transonic Gas Dynamics. Wiley, New York (1958)
8. Busemann, A.: Infinitesimale Kegelige Uberschallstromungenn Schriften. Deutsche Akad. Lufo. **7B**, 105–122 (1943)
9. Cheeger, J.: On the Hodge theory of Riemannian pseudomanifolds. Geometry of the Laplace operator. In: Proc. Sympos. Pure Math., Univ. Hawaii, Honolulu, Hawaii, 1979, vol. 36, pp. 91–146. American Mathematical Society, Providence (1980)
10. Cheeger, J.: Spectral geometry of singular Riemannian spaces. J. Different. Geom. **18**, 575–657 (1983)
11. Darabi, F., Rastkar, A.: A quantum cosmology and discontinuous signature changing classical solutions. Gen. Relat. Gravit. **38**, 1355–1366 (2006)
12. Dereli, T., Tucker, R.W.: Signature dynamics in general relativity. Class. Quantum Grav. **10**, 365–373 (1993)

13. Didenko, V.P.: On the generalized solvability of the Tricomi problem. Ukrain. Math. J. **25**, 10–18 (1973)
14. Dray, T., Ellis, G.F.R., Hellaby, C.: Note on signature change and Colombeau theory. Gen. Relat. Gravit. **33**, 1041–1046 (2001)
15. Dray, T., Ellis, G.F.R., Hellaby, C., Manogue,C.A.: Gravity and signature change. Gen. Relat. Gravit. **29**, 591–597 (1997)
16. Dray, T., Manogue,C.A., Tucker, R.W.: Particle production from signature change. Gen. Relat. Gravit. **23**, 967–971 (1991)
17. Ellis, G., Sumeruk, A., Coule, D., Hellaby, C.: Change of signature in classical relativity Class. Quantum Grav. **9**, 1535–1554 (1992)
18. Embacher, F.: Actions for signature change. Phys. Rev. D **51**, 6764–6777 (1995)
19. Fernando Barbero G., J.: From Euclidean to Lorentzian general relativity: The real way. Phys. Rev. D **54**, 1492–1499 (1996)
20. Gaida, I., Hollman, H.R., Stewart, J.M.: Classical and quantum analysis of repulsive singularities in four-dimensional extended supergravity. Class. Quantum Grav. **16**, 2231–2246 (1999)
21. Gibbons, G.W., Ishibashi, A.: Topology and signature changes in braneworlds. Class. Quantum Grav. **21**, 2919–2935 (2004)
22. Gordon, W.: Zur Lichtfortpflanzung nach der Relativitatstheorie. Ann. Phys. Leipzig **72**, 421–456 (1923)
23. Gu, C.: On some differential equations of mixed type in n-dimensional spaces [in Russian]. Sci. Sin. **14**, 1574–1580 (1965)
24. Gu, C.: A global study of extremal surfaces in 3-dimensional Minkowski space. In: Gu, C-H, Berger, M., Bryant, R. L. (eds.) Differential Geometry and Differential Equations, (Shanghai 1985) Lecture Notes in Mathematics, vol. 1255, pp. 26–33. Springer, Berlin (1985)
25. Gu, C.: The extremal surfaces in the 3-dimensional Minkowski space. Acta Math. Sin. n. s. **1**, 173–180 (1985)
26. Gu, C.: On the mixed partial differential equations in n independent variables, Journées Équat. aux dérivées partielles, pp. 1–2. (1980)
27. Gu, C.: On partial differential equations of mixed type in n independent variables. Commun. Pure Appl. Math. **34**, 333–345 (1981)
28. J. B. Hartle, J.B., Hawking, S.W., Wave function of the universe. Phys. Rev. D **28**, 2960–2975 (1983)
29. Hayashi, N., Hirata, H.: Local existence in time of small solutions to the elliptic-hyperbolicn Davey-Stewartson system in the usual Sobolev space. Proc. Edinburgh Math. Soc. **40**, 563–581 (1997)
30. Hayward, S.A.: Comment on "Failure of standard conservation laws at a classical change of signature," Phys. Rev. D **52**, 7331–7332 (1995)
31. Hayward, S.A.: Signature change in general relativity. Class. Quantum Grav. **9**, 1851–1862 (1992)
32. Hellaby, C., Dray, T.: Failure of standard conservation laws at a classical change of signature. Phys. Rev. D **49**, 5096–5104 (1994)
33. Hellaby, C., Dray, T.: Reply comment: Comparison of approaches to classical signature change. Phys. Rev. D **52**, 7333–7339 (1995)
34. Heidmann, J.: Relativistic Cosmology, An Introduction. Springer, Berlin (1980)
35. Hellaby, C., Sumeruk, A., Ellis, G.F.R.: Classical signature change in the black hole topology. Int. J. Mod.Phys. **D6**, 211–238 (1997)
36. Hua, L.K.: Geometrical theory of partial differential equations. In: Chern, S.S., Wen-tsün, W. (eds.), Proceedings of the 1980 Beijing Symposium on Differential Geometry and Differential Equations, pp. 627–654. Gordon and Breach, New York (1982)
37. Hunter, J.K., Tesdall, A.M.: Weak shock reflection. In: Givoli, D., Grote, M.J., Papanicolaou, G.C. (eds.) A Celebration of Mathematical Modeling: The Joseph B. Keller Anniversary Volume, pp. 93–112. Kluwer Academic Publishers, Dordrecht (2004)
38. Ji, X-H., Chen, D-Q.: Tricomi's problems of non-homogeneous equation of mixed type in real projective plane. In: Chern, S.S., Wen-tsün, W. (eds.), Proceedings of the 1980 Beijing

Symposium on Differential Geometry and Differential Equations, pp. 1257–1271. Gordon and Breach, New York (1982)

39. Katsanis, T.: Numerical techniques for the solution of symmetric positive linear differential equations. Ph.D. thesis, Case Institute of Technology (1967)
40. Katsanis, T.: Numerical solution of Tricomi equation using theory of symmetric positive differential equations. SIAM J. Numer. Anal. **6**, 236–253 (1969)
41. Keller, J.B., Blank, A.: Diffraction and reflection of pulses by wedges and corners. Commun. Pure Appl. Math. **4**, 75–94 (1951)
42. Klein, C.: Binary black hole spacetimes with helical Killing vector. Phys. Rev. D **70**, 124026 (2004)
43. Kossowski, M., Kriele, M.: Signature type change and absolute time in general relativity. Class. Quantum Grav. **10**, 1157–1164 (1993)
44. Kossowski, M., Kriele, M.: The Einstein equation for signature type changing spacetimes. Proc. R. Soc. London Ser. A. **446**, 115–126 (1994)
45. Kriele, M., Martin, J.: Black holes, cosmological singularities and change of signature. Class. Quantum Grav. **12**, 503–511 (1995)
46. Liberati, S., Visser, M., Weinfurtner, S.: Naturalness in an emergent analogue spacetime. Phys. Rev. Lett. **96**, 151301 (2006)
47. Lupo, D., Morawetz, C.S., Payne, K.R.: On closed boundary value problems for equations of mixed elliptic-hyperbolic type. Commun. Pure Appl. Math. **60**, 1319–1348 (2007)
48. Lupo, D., Morawetz, C.S., Payne, K.R.: Erratum: "On closed boundary value problems for equations of mixed elliptic-hyperbolic type," [Commun. Pure Appl. Math. **60**, 1319–1348 (2007)]. Commun. Pure Appl. Math. **61**, 594 (2008)
49. Lupo, D., Payne, K.R.: Critical exponents for semilinear equations of mixed elliptic-hyperbolic and degenerate types. Commun. Pure Appl. Math. **56**, 403–424 (2003)
50. Magnanini, R., Talenti, G.: Approaching a partial differential equation of mixed elliptic-hyperbolic type. In: Anikonov, Yu.E., Bukhageim, A.L., Kabanikhin, S.I., Romanov, V.G. (eds.) Ill-posed and Inverse Problems, pp. 263–276. VSP, Utrecht (2002)
51. Mars, M., Senovilla, J.M.M., Vera, R.: Signature change on the brane. Phys. Rev. Lett. **86**, 4219–4222 (2001)
52. Mars, M., Senovilla, J.M.M., Vera, R.: Lorentzian and signature changing branes. Phys. Rev. D **76**, 044029 (2007)
53. Mars, M., Senovilla, J.M.M., Vera, R.: Is the accelerated expansion evidence of a forthcoming change of signature? Phys. Rev. D **77**, 027501 (2008)
54. Morawetz, C.S.: A weak solution for a system of equations of elliptic-hyperbolic type. Commun. Pure Appl. Math. **11**, 315–331 (1958)
55. Otway, T.H.: Nonlinear Hodge maps. J. Math. Phys. **41**, 5745–5766 (2000)
56. Otway, T.H.: Hodge equations with change of type. Ann. Mat. Pura Appl. **181**, 437–452 (2002)
57. Otway, T.H.: Maps and fields with compressible density. Rend. Sem. Mat. Univ. Padova **111**, 133–159 (2004)
58. Otway, T.H.: Harmonic fields on the extended projective disc and a problem in optics. J. Math. Phys. **46**, 113501 (2005)
59. Otway, T.H.: Erratum: "Harmonic fields on the extended projective disc and a problem in optics," [J. Math. Phys. 46, 113501 (2005)]. J. Math. Phys. **48**, 079901 (2007)
60. Otway, T.H.: Variational equations on mixed Riemannian-Lorentzian metrics. J. Geom. Phys. **58**, 1043–1061 (2008)
61. Page, D.N.: Susskind's challenge to the Hartle-Hawking no-boundary proposal and possible resolutions. J. Cosmol. Astropart. Phys. **01** 004 (2007)
62. Payne, K.R.: Singular metrics and associated conformal groups underlying differential operators of mixed and degenerate types. Ann. Mat. Pura Appl. **184**, 613–625 (2006)
63. Schönberg, M.: Classical theory of the point electron. Phys. Rev. **69**, 211–224 (1946)
64. Stewart, J.M.: Signature change, mixed problems and numerical relativity. Class. Quantum Grav. **18**, 4983–4995 (2001)
65. Stillwell, J.: Geometry of Surfaces. Springer, Berlin (1992)

66. Stillwell, J.: Sources of Hyperbolic Geometry. American Mathematical Society, Providence (1996)
67. Stokes, G.G.: Mathematical and Physical Papers, vol. 5, Appendix. Cambridge University Press, Cambridge (1880–1905)
68. Torre, C.G.: The helically reduced wave equation as a symmetric positive system. J. Math. Phys. **44**, 6223–6232 (2003)
69. Vakili, B., Jalalzadeh, S., Sepangi, H.R.: Classical and quantum spinor cosmology with signature change. J. Cosmol. Astropart. Phys. (2005) 006 doi:10.008/1475-7516/2005/05/006
70. Weinfurtner, S.E.C.: Analog model for an expanding universe. Gen. Relat. Grav. **37**, 1549–1554 (2005)
71. Weinfurtner, S.E.C.: Signature-change events in emergent spacetimes with anisotropic scaling. J. Phys. Conf. Ser. **189**, 012046 (2009)
72. Weinfurtner, S., Liberati, S., Visser, M.: Analogue spacetime based on 2-component Bose-Einstein condensates. In: Quantum Analogues: From Phase Transitions to Black Holes and Cosmology. Lecture Notes in Physics vol. 718, pp. 115–163. Springer, Heidelberg (2007)
73. Weinfurtner, S.E.C., White, A., Visser, M.: Trans-Planckian physics and signature change events in Bose gas hydrodynamics. Phys. Rev. D **76**, 124008 (2007)
74. Xu, Z-F., Xu, Z-Y., Chen, G-L.: Some notes on the Busemann equation [in Chinese]. Adv. in Math. (Beijing) **16**, 81–86 (1987)

Appendix A
Summary and Addenda

There is no new material in the first four sections of Appendix A. In Sects. A.1, A.2, definitions of the four kinds of solutions to the homogeneous Dirichlet problem which are discussed in this course are placed in one section for easy comparison. In Sects. A.3 and A.4, the main results reported in the text concerning the well-posedness of the Dirichlet problem for elliptic–hyperbolic equations of Keldysh type are collected. The last two sections of Appendix A contain new material. In A.5, a comparison is made between what is known about the two canonical classes of equations. Section A.6 contains a brief discussion of a very new method for attacking elliptic–hyperbolic boundary value problems.

A.1 Various Notions of Solution

1. *Distribution solution*: If u is a distribution solution to the equation $Lu = f$, then $u \in L^2(\Omega) \ni \forall \xi \in H_0^1(\Omega; \mathcal{K})$ for which $L^*\xi \in L^2(\Omega)$,

$$\left(u, L^*\xi\right) = \langle f, \xi \rangle,$$

   where $(\ ,\ )$ is the L^2 inner product and $\langle\ ,\ \rangle$ is the Lax duality bracket.
2. *Weak solution* (first alternative): If u is a weak solution to the equation $Lu = f$, then $u \in H_0^1(\Omega, \mathcal{K}) \ni \forall \xi \in H_0^1(\Omega; \mathcal{K})$,

$$\langle Lu, \xi \rangle \equiv - \int\int_{\Omega} \left(\mathcal{K}u_x\xi_x + u_y\xi_y\right) dx dy = \langle f, \xi \rangle \qquad (A.1)$$

   where

$$Lu = [\mathcal{K}u_x]_x + u_{yy} = f \qquad (A.2)$$

   with u vanishing identically on $\partial\Omega$. Again, $\langle\ ,\ \rangle$ is the duality pairing between $H_0^1(\Omega; \mathcal{K})$ and $H^{-1}(\Omega; \mathcal{K})$.

T.H. Otway, *The Dirichlet Problem for Elliptic-Hyperbolic Equations of Keldysh Type*, Lecture Notes in Mathematics 2043, DOI 10.1007/978-3-642-24415-5, © Springer-Verlag Berlin Heidelberg 2012

The equivalence on the extreme left-hand side of (A.1) can be understood by integrating Lu and ξ by parts, using the vanishing of u on the boundary.

If the weight function is not taken to be the type-change function, then it may still be possible to show that a distribution solution lies in $H_0^1(\Omega; k)$ for some $k(x, y)$. This was done in Theorem 5 of [18] for the case $K = x - y^2$ and $k = y^2$. While it may be acceptable to call a solution that lies in L^2 a "distribution solution," calling a solution which lies even in a weighted Sobolev space "distributional" is a bit of a stretch. So perhaps such solutions ought to be called *generalized*. It is not clear that such generalized solutions, even if they lie in weighted H^1, are unique unless $k = K$.

3. *Weak solution* (second alternative): If u is a weak solution to the equation $Lu = f$, then there exists a sequence $u_n \in C_0^\infty(\Omega)$ such that

$$||u_n - u||_{H_0^1(\Omega; \mathcal{K})} \to 0 \text{ and } ||Lu_n - f||_{H^{-1}(\Omega; \mathcal{K})} \to 0 \qquad (A.3)$$

as n tends to infinity.

Definitions 2 and 3 can be shown to be equivalent for L having the special form given by (A.2). Taking $\xi = u$ in (A.1), we obtain

$$-\int\int_\Omega \left(Ku_x^2 + u_y^2 \right) dxdy = (f, u).$$

Because C_0^∞ is dense in $H_0^1(\Omega; \mathcal{K})$, the equivalence of the two definitions follows.

4. *Strong solution*: If $\mathbf{u} \in L^2$ is a strong solution to the equation $L\mathbf{u} = \mathbf{f}$, where $\mathbf{f}$ is a given L^2 vector, with given boundary conditions, then there exists a sequence $\mathbf{u}^\nu$ of continuously differentiable vectors, satisfying the boundary conditions, for which $\mathbf{u}^\nu$ converges to $\mathbf{u}$ in L^2 and $L\mathbf{u}^\nu$ converges to $\mathbf{f}$ in L^2.

Definition 4 is often applied to first-order systems in which $\mathbf{u} = (u_1, u_2)$ and we can take $u_1 = \varphi_x$ and $u_2 = \varphi_y$. In that case, $\mathbf{u} \in L^2$ is equivalent to $\varphi \in H^{1,2}$.

Any of the above definitions can be extended to mixed Dirichlet–Neumann problems – which are discussed in Sects. 3.3 and 3.5 – by imposing Dirichlet and Neumann conditions on disjoint proper subsets of the boundary.

A.2   Comparison of Methods

Suppose that we want to establish a fundamental inequality of the general form (3.34); that is, we want to show that

$$||u||_U \leq C\,||L^*u||_V$$

for a suitable choice of function spaces U and V, and all $u \in C_0^\infty(\Omega)$.

If we try the abc-method, then we will be seeking functions a, b, and c such that, for some $\delta > 0$,

$$\int\int_\Omega \left(au + bu_x + cu_y\right) L^*u \, dxdy \geq \delta \int\int_\Omega |\mathscr{K}|u_x^2 + u_y^2 dxdy.$$

If we find such a, b, and c, then we reason that

$$\int\int_\Omega \left(au + bu_x + cu_y\right) L^*u \, dxdy \leq \|au + bu_x + cu_y\|_{L^2(\Omega)} \|L^*u\|_{L^2(\Omega)}$$

$$\leq C'_{\mathscr{K}} \|u\|_{H_0^1(\Omega;\mathscr{K})} \|L^*u\|_{L^2(\Omega)},$$

where $C'_{\mathscr{K}}$ depends on $\sup\left(|\mathscr{K}/\mathscr{K}'|\right)$. Thus we are forced to choose $U = H_0^1(\Omega;\mathscr{K})$ and $V = L^2(\Omega)$ in the fundamental inequality (3.34). Now we define, as in Sect. 3.4,

$$J_f\left(L^*\xi\right) \equiv \langle f, \xi\rangle \leq \|f\|_{H^{-1}(\Omega;\mathscr{K})} \|\xi\|_{H_0^1(\Omega;\mathscr{K})} \leq C\|f\|_{H^{-1}(\Omega;\mathscr{K})} \|L^*\xi\|_{L^2(\Omega)}. \tag{A.4}$$

We find that if $f \in H^{-1}(\Omega;\mathscr{K})$, then J_f is a bounded function on the subspace of L^2 consisting of elements $\xi \in C_0^\infty(\Omega)$ for which $L^*\xi$ is bounded in L^2. Hahn–Banach arguments extend inequality (A.4) to the L^2-closure and we apply the Riesz Representation Theorem in the inner product space L^2. This allows us to show the existence of a *distribution solution* in L^2.

Now suppose that we prove (3.34) using the integral variant of the abc method, under the hypothesis that Ω is star-shaped with respect to the vector field $-(b,c)$. Defining H as in (4.44) and (4.45) we find, as in (4.57) and (4.58), that

$$\delta \int\int_\Omega \left(|\mathscr{K}|v_x^2 + v_y^2\right) dxdy \leq \left(v, L^*Hv\right) = \left(v, L^*u\right) \leq \|v\|_{H_0^1(\Omega;\mathscr{K})} \|L^*u\|_{H^{-1}(\Omega;\mathscr{K})}.$$

After dividing through by the $H_0^1(\Omega;\mathscr{K})$-norm of v, we find that we can bound the left-hand side of the preceding inequality below by the $L^2(\Omega;|\mathscr{K}|)$-norm of u, using (4.44) and (4.45). Replacing (A.4) by (4.60), we find that

$$J_f\left(L^*\xi\right) \leq C\|f\|_{L^2(\Omega;|\mathscr{K}|^{-1})} \|L^*\xi\|_{H^{-1}(\Omega;\mathscr{K})}.$$

Now applying Hahn–Banach arguments as in the preceding case, we find that we are led by duality to apply the Riesz Representation Theorem in the inner product space $H_0^1(\Omega;\mathscr{K})$ rather than the inner product space L^2. That is, we find that there is a $u \in H_0^1(\Omega;\mathscr{K})$ such that

$$\langle u, L^*\xi\rangle = (f, \xi)_{L^2(\Omega)} \tag{A.5}$$

$\forall \xi \in H_0^1(\Omega; \mathcal{K})$ where $L^* : H_0^1(\Omega; \mathcal{K}) \to H^{-1}(\Omega; \mathcal{K})$ is a unique, continuous extension of the original operator (c.f. (2.8) of [6]). Note that the space weighted-H^{-1} for $L^*\xi$ is appropriate for obtaining u in weighted-H^1, as indicated by the duality bracket on the left-hand side of (A.5). However, due to the form of (A.1), we cannot actually apply this method in an obvious way unless $L = L^*$.

Instead of the L^2 solution which resulted from applying the Riesz Representation Theorem to the abc-method in the previous case, in the present case the Riesz Representation Theorem has given us a solution in weighted-H^1. For that reason, and because definition (A.3) can be used to derive uniqueness, it is appropriate to call the solution *weak*.

So the advantage of the integral variant of the abc method over the conventional abc-method is that the former involves estimates that are one derivative higher than those of the latter method, leading to the application of the Riesz Representation theorem in a higher (although weighted) Sobolev space. Because when applying the Riesz Representation Theorem, $L^*\xi$ is dual to the solution u in the Lax duality bracket, $L^*\xi \in L^2(\Omega)$ implies $u \in L^2(\Omega)$, whereas $L^*\xi \in H^{-1}(\Omega; \mathcal{K})$ implies $u \in H_0^1(\Omega; \mathcal{K})$.

A.3   The Existence of Solutions

In the following, results for inhomogeneous equations having homogeneous boundary conditions possess analogues for homogeneous equations having inhomogeneous boundary conditions via the arguments of Sect. 2.6.

A.3.1   Distribution Solutions

Theorem A.1 (c.f. Theorem 3.3). *Let Ω be a bounded, connected domain having piecewise C^1 boundary. The Dirichlet problem for the equation*

$$\mathcal{K}(x)u_{xx} + u_{yy} + \mathcal{K}'(x)u_x = f(x, y) \tag{A.6}$$

with boundary condition

$$u(x, y) = 0 \ \forall (x, y) \in \partial\Omega$$

possesses a distribution solution $u \in L^2(\Omega)$ for every $f \in H^{-1}(\Omega; \mathcal{K})$, provided $\mathcal{K} = x$.

Remark. See also [18] for the cold-plasma case, [15] for a case related to the Monge–Ampère equation, and [6] for the original method introduced in the context of Tricomi-type equations.

A.3.2   Weak Solutions

Theorem A.2 (c.f. Theorem 4.2). *There exists a unique, weak solution to the Dirichlet problem of Theorem A.1 provided the type-change function $\mathscr{K}(x)$ is replaced by the type-change function $\mathscr{K}(x, y) = x - y^2$, and in addition the following hypotheses are satisfied: x is non-negative on Ω and the origin of coordinates lies on $\partial\Omega$; Ω is star-shaped with respect to the flow of the vector field $V = -(b, c)$ for $b = mx$ and $c = \mu y$, where m and μ are positive constants and m exceeds 3μ.*

Theorem A.3 (c.f. Theorem 4.3). *Consider an equation having the form*

$$Lu \equiv x^{2k+1}u_{xx} + u_{yy} + c_1 x^{2k} u_x + c_2 u = 0, \tag{A.7}$$

where $k \in \mathbf{Z}^+$ and the constants c_1 and c_2 satisfy $c_1 < k + 1$ and $c_2 < 0$ with $|c_2|$ sufficiently large. Let a portion of the line $x = 0$ lie in Ω and let the point $(0, 0)$ lie on $\partial\Omega$. Assume that Ω is star-shaped with respect to the vector field $V = -(b, c)$, where $b = mx$, and $c = \mu y$. Let μ be a positive constant and let

$$m = \begin{cases} -a/\ell + \mu/2\ell - \delta/\ell \text{ in } \Omega^+ \\ -a/\ell + \mu/2\ell + \delta/\ell \text{ in } \Omega^- \end{cases}$$

for a positive constant δ, where $\ell = k + 1 - c_1$. Let a be a negative constant of sufficiently large magnitude. In particular, let a have sufficiently large magnitude that m is positive. Then for every $f \in L^2\left(\Omega; |\mathscr{K}|^{-1}\right)$ there is a distribution solution u to (A.7), with

$$u(x, y) = 0 \ \forall (x, y) \in \partial\Omega,$$

lying in $H_0^1(\Omega; \mathscr{K})$ for $\mathscr{K} = x^{2k+1}$.

Remark. For the original Tricomi case of the preceding two results, see [6].

Theorem A.4 (Magnanini–Talenti [14] (c.f. Theorem 5.1)). *Let D_R, $R > 1$, be a disc of radius R centered at the origin of coordinates in $\mathbf{R}^2$ and let Ω be a subset of D_R that has positive distance from both the boundary and the center of D_R. Suppose that $f \in L^2[-\pi, \pi]$ is a given function. Choose polar coordinates (r, θ), where $0 < r < \infty$, $-\pi < \theta \leq \pi$. Then there is a unique function V lying in $H^{1,2}(\Omega)$, which satisfies*

$$\left(r^2 - 1\right) V_{rr} + r V_r + V_{\theta\theta} = 0$$

weakly in D_R and smoothly in the punctured disc D_r, where $0 < r < R$. Moreover, $V = f$ on ∂D in the sense that

$$\lim_{r\uparrow R} \int_{-\pi}^{\pi} |V\left(r \cdot e^{i\theta}\right) - f(\theta)|^2 d\theta = 0.$$

Finally, V has an explicit representation in the form of a series

$$V\left(re^{i\theta}\right) = \frac{a_0}{2} + \sum_{k=1}^{\infty} \frac{T_k(r)}{T_k(R)} \cdot \{a_k \cos(k\theta) + b_k \sin(k\theta)\},$$

where

$$a_k = \pi^{-1} \int_{-\pi}^{\pi} f(\theta) \cos(k\theta)\, d\theta;$$

$$b_k = \pi^{-1} \int_{-\pi}^{\pi} f(\theta) \sin(k\theta)\, d\theta;$$

$$T_k(r) = \begin{cases} \cos(k \arccos r) & \text{if } r \le 1, \\ \frac{1}{2}\left[\left(r + \sqrt{r^2 - 1}\right)^k + \left(r - \sqrt{r^2 - 1}\right)^k\right] & \text{if } r > 1. \end{cases}$$

Theorem A.5 (c.f. Theorem 6.3). *Denote by Ω a region of the plane for which part of the boundary $\partial\Omega$ consists of a family $\mathcal{G}$ of curves composed of points satisfying*

$$\left(1 - y^2\right)dx^2 + 2xy\,dx\,dy + (1 - x^2)dy^2 = 0$$

and the remainder $C = \partial\Omega \setminus \mathcal{G}$ of the boundary consists of points (x, y) which do not satisfy that equation. In addition to the requirements that Ω contain an arc of the unit circle and that y^2 is bounded above away from unity, we require that any points satisfying the equation

$$xy + 1 - y^2 = 0$$

lie in the complement of Ω. Let $\partial\Omega$ be oriented in the counter-clockwise direction and have piecewise continuous tangent. Let the parameter θ lie in the interval $[0, \pi/4]$ and denote by Ω the region of the first and fourth quadrants bounded by the characteristic line

$$\mathcal{G}_1 : x \cos\theta + y \sin\theta = 1,$$

the characteristic line

$$\mathcal{G}_2 : x \cos\theta - y \sin\theta = 1,$$

and a smooth curve C. Let C intersect the lines $\mathcal{G}_1, \mathcal{G}_2$ at two distinct points c_1, c_2, respectively. Assume that $dy \le 0$ on C if γ_1 and γ_2 are negative definite and that $dy \ge 0$ if γ_1 and γ_2 are positive definite. Suppose that the bounded functions γ_1 and γ_2 are definite with the same sign. Then there exists a weak solution of the boundary value problem

$$\mathbf{L}\mathbf{u} = \mathbf{f} \ in \ \Omega;$$

$$(\mathbf{L}\mathbf{u})_1 = \left[\left(1-x^2\right)u_1\right]_x - 2xyu_{1y} + \left[\left(1-y^2\right)u_2\right]_y + \gamma_2 u_1,$$

$$(\mathbf{L}\mathbf{u})_2 = \left(1-y^2\right)\left(u_{1y} - u_{2x}\right) + \gamma_1 u_2,$$

where

$$\gamma_i = \left(xy + 1 - y^2\right)\Gamma_i, \ i = 1, 2;$$

$$u_1\frac{dx}{ds} + u_2\frac{dy}{ds} = 0 \ on \ \partial\Omega$$

for every $\mathbf{f}$ in the space H consisting of pairs of measurable functions $\mathbf{h} = (h_1, h_2)$ for which the norm

$$\|\mathbf{h}\|^* = \left[\int\int_\Omega \left(|\gamma_1\,(x, y)|^{-1}\,h_1^2 + |\gamma_2\,(x, y)|^{-1}\,h_2^2\right)dxdy\right]^{1/2}$$

is finite.

A.3.3 Strong Solutions

Theorem A.6 (c.f. Theorem 3.4). *Let Ω be a bounded, connected domain of $\mathbf{R}^2$ having C^2 boundary $\partial\Omega$, oriented in a counterclockwise direction. Let $\partial\Omega_1^+$ be a (possibly empty and not necessarily proper) subset of $\partial\Omega^+$. Let the inequality*

$$bn_1 + cn_2 \geq 0$$

be satisfied on $\partial\Omega^+ \backslash \partial\Omega_1^+$. On $\partial\Omega_1^+$ let

$$bn_1 + cn_2 \leq 0$$

and on $\partial\Omega \backslash \partial\Omega^+$, let

$$-bn_1 + cn_2 \geq 0.$$

Let $b(x, y)$ and $c(x, y)$ satisfy

$$b^2 + c^2 \mathscr{K} \neq 0$$

on Ω, with neither b nor c vanishing on Ω^+, and let

$$\mathscr{K}\,(bn_1 - cn_2)^2 + (c\mathscr{K}n_1 + bn_2)^2 \leq 0 \ on \ \partial\Omega \backslash \partial\Omega^+.$$

Let L be given by the system

$$Lu = f, \tag{A.8}$$

$$Lu = \begin{pmatrix} \mathcal{H}(x,y) & 0 \\ 0 & -1 \end{pmatrix} \begin{pmatrix} u_1 \\ u_2 \end{pmatrix}_x + \begin{pmatrix} 0 & 1 \\ 1 & 0 \end{pmatrix} \begin{pmatrix} u_1 \\ u_2 \end{pmatrix}_y$$

$$+ \text{ zeroth-order terms} \tag{A.9}$$

and let EL be symmetric positive, where

$$E = \begin{pmatrix} b & -c\mathcal{H} \\ c & b \end{pmatrix}.$$

Let the Dirichlet condition

$$-u_1 n_2 + u_2 n_1 = 0$$

be satisfied on $\partial\Omega^+ \backslash \partial\Omega_1^+$ and let the Neumann condition

$$\mathcal{H} u_1 n_1 + u_2 n_2 = 0$$

be satisfied on $\partial\Omega_1^+$. Then the equations (A.8), (A.9) possess a strong solution on Ω for every $f \in L^2(\Omega)$. If in particular, the operator L is given by

$$(Lu)_1 = xu_{1x} + u_{2y} + \kappa_1 u_1 + \kappa_2 u_2,$$

$$(Lu)_2 = u_{1y} - u_{2x},$$

where κ_1 and κ_2 are constants, then sufficient conditions for the system to be symmetric positive are

$$2b\kappa_1 - b_x\mathcal{H} - b + c_y\mathcal{H} > 0 \text{ in } \Omega$$

and

$$\left(2b\kappa_1 - b_x\mathcal{H} - b + c_y\mathcal{H} \right) \left(2c\kappa_2 + b_x - c_y \right)$$

$$- \left(b\kappa_2 + c\kappa_1 - c_x\mathcal{H} - c - b_y \right)^2 > 0 \text{ in } \Omega.$$

Theorem A.7 (c.f. Theorem 5.2). *Consider a system having the form*

$$Lw = A^1 w_r + A^2 w_\theta + Bw = F,$$

where L is a first-order operator, $w = (w_1(r,\theta), w_2(r,\theta))$, $F = (f,0)$,

$$A^1 = \begin{pmatrix} \mathcal{H}(r) & 0 \\ 0 & -1 \end{pmatrix}, \quad A^2 = \begin{pmatrix} 0 & 1 \\ 1 & 0 \end{pmatrix},$$

and

$$B = \begin{pmatrix} \mathcal{K}'(r) \, k \\ 0 \quad 0 \end{pmatrix}.$$

where $\mathcal{K}'(r) > 0$; k is a nonzero constant; $\mathcal{K}(r) < 0$ for $0 \leq r < r_{crit}$ and $\mathcal{K}(r) > 0$ for $r_{crit} < r \leq R$. Suppose that there is a positive constant v_0 such that $\mathcal{K}'(r) \geq v_0$. Let there be continuous functions $\sigma(\theta)$ and $\tau(\theta)$ such that the boundary condition

$$\sigma(\theta)w_1 + \tau(\theta)w_2 = 0$$

is satisfied on the boundary $r = R$, where the product $\sigma(\theta)\tau(\theta)$ is either strictly positive or strictly negative and has sign opposite to the sign of k. Then the resulting boundary value problem possesses a strong solution on the closed disc $\{(r, \theta) \, | 0 \leq r \leq R\}$ provided $|\mathcal{K}(0)|$ is sufficiently small.

A.3.4   Classical Solutions

Theorem A.8 (Cinquini-Cibrario [1] (c.f. Sect. 3.6.2)). *Given the domain bounded in the half-plane $x \geq 0$ by the curve*

$$C = \{(x, y) \, | 4x + y^2 = 1\}$$

and in the half-plane $x < 0$ by the intersecting characteristic lines

$$\Gamma^{\pm} = \left\{(x, y) \, | y = \pm 2 \left(\sqrt{-x} - 2\right)\right\},$$

and the equation

$$xu_{yy} + u_{yy} = 0,$$

then a solution to this equation exists which is equal on C to a function of the form $x\varphi(y)$, where $\varphi(y)$ is finite and continuous on C, including the endpoints, and vanishes on the y-axis. The solution is analytic in the interior of the domain.

Theorem A.9 (Gu [2, 3] (c.f. Theorem 6.4)). *For $s > 1$, let $f \in H^{1,s}(\Omega)$ and $a > -(n/2)+s$. Then there exists a unique solution $\varphi \in H^{1,s+1}(\Omega)$ to the equation*

$$L\varphi \equiv \left(\delta^{ij} - x^i x^j\right) \partial_{ij}\varphi + 2ax^i \partial_i \varphi - a(a+1)\varphi = f, \qquad (A.10)$$

which satisfies the boundary conditions

$$\varphi_{|\partial\Omega} = 0, \quad \frac{\partial\varphi}{\partial n}\bigg|_{\partial\Omega} = 0; \qquad (A.11)$$

here $\partial/\partial n$ indicates differentiation in the direction of the outward-pointing normal. In the special case $s \geq (n/2) + 2$, a classical solution of (A.10), (A.11) exists. In addition, under the conditions of the theorem the solution is unique whenever $a < -n/2$.

A.4   The Nonexistence of Solutions

Theorem A.10 (c.f. Theorem 3.1). *Consider the equation*

$$\mathscr{K}(x)u_{xx} + u_{yy} + \frac{\mathscr{K}'(x)}{2}u_x = 0,$$

where $\mathscr{K}$ satisfies $\mathscr{K}(0) = 0$ and $x\mathscr{K}(x) > 0$ for $x \neq 0$. Assume that $\mathscr{K}$ is C^1, and monotonic on $\mathscr{D}^-$. Define constants a, b, d, and m, where $m < a \leq 0 < d$ and $b > 0$. Consider the domain $\mathscr{D}$ formed by the line segments

$$\mathscr{L}_1 = \{(x,y)\,|a \leq x \leq d, y = -b\};$$

$$\mathscr{L}_2 = \{(x,y)\,|x = d, -b \leq y \leq b\};$$

$$\mathscr{L}_3 = \{(x,y)\,|a \leq x \leq d, y = b\};$$

the characteristic line Γ_1 joining the points $(m,0)$ and $(a,-b)$; and the characteristic line Γ_2 joining the points $(m,0)$ and (a,b). Let the solution u be sufficiently smooth so that the integral

$$I = \int_0^{(x,y)} \left[\mathscr{K}(x)u_x^2 - u_y^2\right]dy - 2u_xu_y dx$$

is continuous on $\mathscr{D} \cup \partial\mathscr{D}$. If u vanishes identically on the non-characteristic boundary, then $u \equiv 0$ on all of $\mathscr{D}$. Consequently, the closed Dirichlet problem on $\mathscr{D}$ is over-determined.

Remark. For the original Tricomi case, see [16], Theorem 2. The conclusions of the theorem extend to certain equations having type-change functions which do not precisely satisfy the conditions satisfied by $\mathscr{K}$, on domains which are not identical to $\mathscr{D}$. These include equations for the Laplace–Beltrami operator on extended $\mathbf{P}^2$ (Theorem 6.1) and on a relativistically rotating disc (Problem 11, Appendix B); an equation arising in non-geometrical optics (Theorem 6.2), and an equation characterizing the behavior of electromagnetic waves in zero-temperature plasma – this last result is due to Morawetz, Stevens, and Weitzner [17]. It is also possible to extend this theorem to certain mixed Dirichlet–Neumann problems (Theorem 3.2).

A.5   Rough Comparison of Tricomi and Keldysh Classes

Recall that there are at least three fundamental differences between the analytic properties of equations of Tricomi type and those of Keldysh type:

1. Characteristic lines associated with equations of Keldysh type degenerate at the parabolic transition, in that they intersect the sonic curve tangentially.
2. Differential operators of Tricomi type tend to be of real principal type, whereas we do not expect operators of Keldysh type to possess this property.
3. Equations of Tricomi type are formally self-adjoint in their second-order terms, whereas equations of Keldysh type require the addition of a suitable first-order term in order to become formally self-adjoint.

As a result of these three differences, there is a significant difference between what we know about the two classes. Here we list five typical differences. The list is by no means exhaustive, and reflects the interests of the text. It is, in that sense, a motivation for Appendix B on suggested directions for future research.

1. Lupo, Morawetz and Payne [6] have established the existence of a unique H^1_{loc} weak solution to the homogeneous closed Dirichlet problem for an inhomogeneous Tricomi equation of the form (1.1). Moreover, these authors showed that if the inhomogeneous term f satisfies $f_x \in L^2\left(\Omega, |y|^{-1}\right)$, then the solution lies in $H^2_{loc}(\Omega)$. In that case, u is continuous by the Sobolev Theorem. It is known that the corresponding weak solution for the formally self-adjoint form of the cold plasma model (Chap. 4) – which is an equation of Keldysh type having somewhat similar structure – is not $H^{1,2}$ in any neighborhood of the origin [17]. See item *vii)* of Appendix B, Problem 6.
2. A maximum principle has been proven by Lupo and Payne for generalized solutions to the *Tricomi problem* (solutions prescribed on the elliptic part of the boundary and a characteristic) [11]. One would expect some kind of extension of this result to the Keldysh case, but no such extension appears to exist in the literature. See Problem 9 of Appendix B.
3. Lupo and Payne have proven a series of results on open boundary value problems for semilinear Tricomi operators [7], [9–13]. These authors have, in particular, investigated a variational approach to semilinear equations of Tricomi type [9]. They have also studied, with occasional collaborators, the spectral properties of linear and semilinear Tricomi operators [5, 8, 12]. I know of no such results for elliptic–hyperbolic operators of Keldysh type. See Problems 15, 12, and 18 of Appendix B.
4. As was indicated earlier, there is a microlocal theory for equations of Tricomi type which has no obvious analogue for equations of Keldysh type. This results in technical information about the solvability of boundary value problems and the propagation of singularities which is lacking for equations of Keldysh type; see [19] and references therein.
5. An existence theorem for the closed Neumann problem has been proven for the Lavrent'ev–Bitsadze equation (2.5) by Pilant [20]. No other result for conormal

conditions on the entire boundary is known for elliptic–hyperbolic equations of either type. See Problem 13 of Appendix B. Some results have been proven by Lupo, Morawetz, and Payne [6] for the mixed Dirichlet–Neumann problem for systems corresponding to equations having the Tricomi form (3.6), (3.7), (3.8); these extend easily to certain equations having the corresponding Keldysh form (3.7), (3.10) [18].

A.6   Weak Solutions in Anisotropic Sobolev Spaces (*After M. Khuri*)

We outline a very recent method introduced by M. Khuri [4], which has not yet been published. For details, see arXiv:1106.4000v1.

Define the Sobolev space $H^{(m,\ell)}(\Omega)$ to consist of functions for which derivatives up to the mth partial derivative in x, and the ℓth partial derivative in y, are square-integrable. We have the norms

$$||u||^2_{(m,\ell)} \equiv \int_\Omega \sum_{0\leq s\leq m, 0\leq t\leq \ell} \left(\partial_x^s \partial_y^t u\right)^2$$

and

$$||v||_{(-m,-\ell)} = \sup_{u\in H^{(m,\ell)}(\Omega)} \frac{|(u,v)|}{||u||_{(m,\ell)}}.$$

We obtain the rigged triple

$$H^{(m,\ell)}(\Omega) \subset L^2(\Omega) \subset H^{(-m,-\ell)}(\Omega).$$

Consider the boundary value problem

$$Lu \equiv Ku_{xx} + u_{yy} + Au_x + Bu_y = f \text{ in } \Omega \qquad (A.12)$$

$$Bu \equiv \alpha u_x + \beta u_y + \gamma u = 0 \text{ on } \partial\Omega, \qquad (A.13)$$

and the adjoint problem

$$L^*v = g \text{ in } \Omega, \qquad (A.14)$$

$$B^*v = 0 \text{ on } \partial\Omega, \qquad (A.15)$$

where as usual the superscripted asterisk denotes formal adjoint. The coefficients K, A, B, α, β, and γ are assumed to be sufficiently smooth on $\overline{\Omega}$. We say that u is a *weak solution* of the boundary value problem (A.12), (A.13) if

$$\left(u, L^*v\right) = (f, v) \quad \forall v \in C^\infty_{B^*}(\overline{\Omega}),$$

where $C_{B^*}^\infty(\overline{\Omega})$ is the space of smooth functions up to the boundary on Ω which satisfy the boundary conditions (A.15).

Theorem A.11 (Khuri [4]). *Let $m, \ell, s, t \in \mathbf{Z}^+ \cup \{0\}$. There exists a weak solution $u \in H^{(m,\ell)}(\Omega)$ to the boundary value problem (A.12), (A.13) if and only if there exists a constant C such that*

$$||v||_{(-s,-t)} \le C ||L^* v||_{(-m,-\ell)} \quad \forall v \in C_{B^*}^\infty(\overline{\Omega}). \tag{A.16}$$

For a proof we refer the reader to the appendix to Khuri's paper.

In order to apply this result, one proceeds in a roughly analogous manner to the case of weighted Sobolev spaces discussed in Chap. 4. Initially, consider the auxiliary boundary value problem

$$Mu = v \text{ in } \Omega,$$

$$\tilde{B}u = 0 \text{ on } \partial\Omega,$$

where the differential operator M and the boundary operator $\tilde{B}$ must be chosen, based on the conditions of the original problem. Apply integration by parts to obtain

$$\left(L^* v, u\right) - (v, Lu) = \int_{\partial\Omega} I_1(u, v)$$

and

$$(Mu, Lu) = \int_\Omega I_2(u, u) + \int_{\partial\Omega} I_3(u, u)$$

for quadratic forms I_1, I_2, and I_3. The method requires that M, $\tilde{B}$, and B^* be chosen so that the following three inequalities are satisfied:

$$\int_\Omega I_2(u, u) \ge C^{-1} ||u||_{(m,\ell)}^2,$$

$$||v||_{(-s,-t)} \le C ||u||_{(m,\ell)}, \tag{A.17}$$

and

$$\int_{\partial\Omega} [I_1(u, Mu) + I_3(u, u)] \ge 0. \tag{A.18}$$

If these three inequalities can be established, then one reasons as follows:

$$||u||_{(m,\ell)} ||L^* v||_{(-m,-\ell)} \ge \left(L^* v, u\right)$$

$$= (v, Lu) + \int_{\partial\Omega} I_1(u, v)$$

$$= (Mu, Lu) + \int_{\partial\Omega} I_1(u, v)$$

$$= \int_{\Omega} I_2\,(u,u) + \int_{\partial\Omega} \left[I_1\,(u,Mu) + I_3\,(u,u)\right]$$

$$\geq C^{-1} ||u||^2_{(m,\ell)}.$$

Dividing through by the $H^{(m,\ell)}$-norm of u and then applying (A.17), one obtains (A.16). Then the weak existence of solutions to the original problem (A.12), (A.13) will follow from Theorem A.11.

References

1. Cibrario, M.: Intorno ad una equazione lineare alle derivate parziali del secondo ordine di tipe misto iperbolico-ellittica. Ann. Sc. Norm. Sup. Pisa, Cl. Sci., Ser. 2. **3**(3, 4), 255–285 (1934)
2. Gu, C.: On the mixed partial differential equations in n independent variables, Journées Équations aux dérivées partielles, pp. 1–2 (1980)
3. Gu, C.: On partial differential equations of mixed type in n independent variables. Commun. Pure Appl. Math. **34**, 333–345 (1981)
4. Khuri, M.A.: Boundary value problems for mixed type equations and applications. J. Nonlinear Anal. Ser. A: TMA (to appear); arXiv:1106.4000v1
5. Lupo, D., Micheletti, A.M., Payne, K.R.: Existence of eigenvalues for reflected Tricomi operators and applications to multiplicity of solutions for sublinear and asymptotically linear nonlical Tricomi problems. Adv. Differ. Equat. **4**, 391–412 (1999)
6. Lupo, D., Morawetz, C.S., Payne, K.R.: On closed boundary value problems for equations of mixed elliptic-hyperbolic type. Commun. Pure Appl. Math. **60**, 1319–1348 (2007)
7. Lupo, D., Payne, K.R.: Multiplicity of nontrivial soutions for an asymptotically linear nonlocal Tricomi problem. Nonlinear Anal. Ser. A: Theor. Meth. **46**, 591–600 (2001)
8. Lupo, D., Payne, K.R.: Existence of a principal eignevalue for the Tricomi problem. Proceedings of the Conference on Nonlinear Differential Equations (Coral Gables, Fl., 1999), pp. 173–180, Electronic J. Differential Equations Conf. **5**, Southwest Texas State University, San Marcos, TX (2000)
9. Lupo, D., Payne, K.R.: A dual variational approach to a class of nonlocal semilinear Tricomi problems. Nonlinear Differ. Equat. Appl. **6** 247–266 (1999)
10. Lupo, D., Payne, K.R.: The dual variational method in nonlocal semilinear Tricomi problems. Nonlinear Analysis and its Applications to Differential Equations (Lisbon, 1998) Progr. Nonlinear Diff Equations Appl., vol. 43, pp. 321–338. Birkhäuser Boston, Boston (2001)
11. Lupo, D., Payne, K.R.: On the maximum principle for generalized solutions to the Tricomi problem. Commun. Contemp. Math. **2** 535–557 (2000)
12. Lupo, D., Payne, K.R.: Spectral bounds Tricomi problems and application to semilear existence and existence with uniqueness results. J. Differ. Equat. **184**, 139–162 (2002)
13. Lupo, D., Payne, K.R.: Critical exponents for semilinear equations of mixed elliptic-hyperbolic and degenerate types. Commun. Pure Appl. Math. **56**, 403–424 (2003)
14. Magnanini, R., Talenti, G.: Approaching a partial differential equation of mixed elliptic-hyperbolic type. In: Anikonov, Yu.E., Bukhageim, A.L., Kabanikhin, S.I., Romanov, V.G. (eds.) Ill-posed and Inverse Problems, pp. 263–276. VSP, Utrecht (2002)
15. Xu, M., Yang, X-P.: Existence of distributional solutions of closed Dirichlet problem for an elliptic-hyperbolic equation. J. Nonlinear Analysis Ser. A: TMA, to appear
16. Morawetz, C.S.: Note on a maximum principle and a uniqueness theorem for an elliptic-hyperbolic equation. Proc. R. Soc. London, Ser. A **236**, 141–144 (1956)

17. Morawetz, C.S., Stevens, D.C., Weitzner, H.: A numerical experiment on a second-order partial differential equation of mixed type. Commun. Pure Appl. Math. **44**, 1091–1106 (1991)
18. Otway, T.H.: Energy inequalities for a model of wave propagation in cold plasma. Publ. Mat. **52**, 195–234 (2008)
19. Payne, K.R.: Propagation of singularities for solutions to the Dirichlet problem for equations of Tricomi type, Rend. Sem. Mat. Univ. Pol. Torino **54**, 115–137 (1996)
20. Pilant, M.: The Neumann problem for an equation of Lavrent'ev-Bitsadze type. J. Math. Anal. Appl. **106**, 321–359 (1985)

Appendix B
Directions for Future Research

In this intentionally speculative appendix, 21 possible lines of investigation for equations of the general form studied in this review are discussed. These include both research into properties of the equations themselves and possible areas of application.

1. With very few exceptions (e.g., [35, 130]), virtually nothing about the existence of solutions to the equations considered in these notes is known for dimensions exceeding two. The situation is somewhat better for equations of Tricomi type, at least since the famous paper by Protter in 1954 [101]. See [122] for a recent application and, e.g., [43] for an earlier higher-dimensional paper. In addition, there are higher-dimensional results using microlocal methods (see, e.g., [89, 90]), and series of technical papers by Karatoprakliev [44–48] and Sorokina [111–116]; see also [51]. Friedrichs' theory of symmetric positive operators is an n-dimensional argument, and although it is often applied in dimension 2, it has been employed in its full n-dimensional generality in many cases involving equations of Tricomi type; see, e.g., [44, 46, 79, 100, 111–114, 124]. One expects that some of these arguments are extendable to equations of Keldysh type.

One of the obstructions to higher-dimensional results for elliptic–hyperbolic equations is the linearization technique – most of the equations originate in non-linear models. The requirements for the hodograph method, for example, are very restrictive. Another obstruction appears to result from the nature of the multiplier methods, which convert the equation to a form in which an energy inequality such as (4.55) can be applied. These methods (that is, the Friedrichs abc method) are easiest to apply when the matrix is 2×2; the abc method can be quite difficult to apply even in the 2×2 case. Finally, underlying many of the current methods for equations of mixed type is the theory of complex variables; see, e.g., the application of the abc method by Morawetz in [76].

Due to the importance of ideas from complex variables in classical elliptic–hyperbolic analysis, it is possible that the three-dimensional, and even four-dimensional cases could be attacked using the quaternion representation. Wu uses quaternions to represent the three-dimensional case of the (hyperbolic) water wave equations in [129]. She does this in order to be able to use a Cauchy integral formula for the 3-D problem, analogously to the use of the Poisson integral formula for the 2-D problem. It might be natural, in a model of rotational compressible flow, to represent the velocity field as a quaternion-valued 1-form. It is even possible that the 2-form representing the vorticity of the flow can be given an interpretation as the curvature associated to the quaternion-valued connection 1-form. If so, the invariance of the vorticity under galilean transformations of the velocity would inherit a geometrical interpretation. Notice that, for an irrotational flow, the vorticity is zero and we are in the flat, geometrically trivial case. See, e.g., [42] and Sect. 1.5 of [7] for examples of this kind of thinking in other contexts. The continuity equations for such a model might turn out to be bundle-valued elliptic–hyperbolic equations having an immediate extension to $\mathbf{R}^n$; c.f. (4.3) of [72].

There are of course many four-dimensional applications of elliptic–hyperbolic equations (Sect. 6.4.5), which have been pursued in physical contexts.

2. Similar remarks can be made about higher-order equations, although in that case it must be said that compelling physical examples seem to be lacking, which is certainly not the case for higher-dimensional second-order problems. Boundary value problems for higher-order elliptic–hyperbolic equations are treated in [87,88]; see the references in [87] for a few other examples, but the literature on this topic seems to be very small.

The few results that exist demonstrate the richness of higher-order equations. The equation considered in [87] (see also [86]) is n-dimensional and depends on several parameters. Depending on the choice of those parameters, the equation is either of mixed hyperbolic–parabolic or mixed elliptic–parabolic type. A concrete example is provided by the hyperbolic–parabolic equation

$$[\sin(\pi t) - 1]\,\partial_t^6 u(x, y, t) + A\partial_t^5 u + \partial_x^6 u(x, y, t)$$
$$+\partial_y^6 u(x, y, t) + [\cos(\pi t) - C]\,u(x, y, t) = f(x, y, t), \qquad (B.1)$$

were A is a constant; C is a positive constant; $t \in (0, 1)$; $(x, y) \in \mathscr{D}$; where

$$\mathscr{D} = \{x, y \,|\, x^2 + y^2 < R\} \,;$$

$R = $ const. > 0; f is a prescribed L^2-function on $\mathscr{D} \times (0, 1)$. The boundary conditions are

$$\frac{\partial^{|\alpha|}}{\partial_x^{\alpha_1} \partial_y^{\alpha_2}} u(t, x, y)|_\Gamma = 0 \qquad (B.2)$$

for $|\alpha| = \alpha_1 + \alpha_2 \le 2$, where $\Gamma = \partial\mathscr{D} \times (0, 1)$;

$$\partial_t^i u(1, x, y) = \frac{1}{2}\partial_t^i u(0, x, y), \qquad (B.3)$$

where $i = 0, 1, 2, 3, 4, 5$, for all $(x, y) \in \overline{\mathscr{D}}$.

It can be shown that if the constants A and C are sufficiently large, then a unique solution to this boundary value problem exists in the generalized sense that

$$\left(u, L^* v\right) = (f, v) \quad \forall v \in \tilde{C}_*^\infty(\overline{G}),$$

where the parentheses denote L^2-inner product; L^* is the formal adjoint of the differential operator of (B.1); $C_*^\infty(\overline{G})$ is the space of infinitely smooth functions satisfying boundary conditions adjoint to (B.2), (B.3); $\overline{G}$ is the closure of the domain $G \equiv \mathscr{D} \times (0, T)$ for $T > 0$. The solution u is an element of the anisotropic function space $H_{(t,x,y)}^{5,3}(G)$. The space $H_{(t,x,y)}^{p,q}(G)$ is defined as the closure of the space of infinitely smooth functions on $\overline{G}$ which satisfy the boundary conditions (B.2), (B.3) with respect to the norm

$$||u||_{p,q}^2 = \int_G \sum_{qi+p|\alpha|\leq pq} \left[\partial_t^i \frac{\partial^{|\alpha|}}{\partial_x^{\alpha_1} \partial_y^{\alpha_2}} u\,(t, x, y)\right]^2 dt\,dx\,dy.$$

The proof uses the ideas of Berezanskii and his school, as outlined in Chaps. 3 and 4, in the context of anisotropic function spaces. See also item $ii)$ of Problem 4, and Sect. A.6. Does this example suggest a higher-order analogue for the method described in Sect. A.6?

3. With a few exceptions, there do not seem to be results in the literature for equations of Keldysh type having multiple parabolic transitions, although such results have been known for Tricomi-type equations for a long time (see, e.g., [106, 113] for early examples; [102] for a later example; and [55, 103] for recent examples).

Multiple parabolic transitions are, in fact, a feature of one of the earliest treatments of elliptic–hyperbolic equations. In 1929, Bateman [12] considered the equation

$$\left(1 - x^2\right) u_{xx} - 2xu_x - \left(1 - y^2\right) u_{yy} + 2yu_y = 0. \tag{B.4}$$

This equation, which can be seen to be of Keldysh type by expressing it in polar coordinates, is hyperbolic when $1 - x^2$ and $1 - y^2$ have the same sign and elliptic when they have different signs. Bateman found some solutions in terms of special functions and made some observations about the apparent lack of uniqueness in certain boundary value problems; but as far as I know there are no formal results for the corresponding boundary value problem in the literature.

Much later, but nonetheless more than 30 years ago, Gu [35] considered the class of n-dimensional equations

$$\left[e(r^2)\delta^{ij} - x^i x^j\right] u_{x^i x^j} + 2ax^i u_{x^j} - a\,(a + 1)\,u = f,$$

where $r^2 = \left(x^1\right)^2 + \cdots (x^n)^2$ and $e\left(r^2\right)$ is a smooth function such that $e\left(r^2\right) - r^2$ is:

B   Directions for Future Research

positive if $0 \leq r^2 < c_1, c_2 < r^2 < c_3$;

negative if $c_1 < r^2 < c_2, r^2 > 5$;

zero if $c_4 < r^2 < c_5$;

here $c_1, \ldots, c_5$ are constants. Taking the domain to be a closed and bounded region containing the sphere $r^2 = c_5$ and assuming that the tangent planes to the boundary of the domain do not meet this sphere, there are two elliptic regions ($0 \leq r^2 < c_1$, $c_2 < r^2 < c_3$), two hyperbolic regions ($c_2 < r^2 < c_1, r^2 > 5$), and one parabolic region ($c_4 < r^2 < c_5$). Gu proves that if a is sufficiently large, then the equation admits a C^2 solution satisfying the conditions

$$ u = 0, \quad \partial u / \partial n = 0 $$

on the domain boundary, where $\partial / \partial n$ indicates differentiation in the direction of the outward-pointing normal. Moreover, if $-a$ is sufficiently large, then there is only one classical or weak solution to this boundary value problem.

Gu's arguments are based on the theory that he developed for the equations considered in Sects. 6.4.2 and 6.4.3. As in the case of (6.36), in some sense (B.4) is a generalization of an equation treated in this review – in this case, (5.53). But again, Gu's method relies on hypotheses which explicitly exclude that case. Nevertheless, one could modify a number of equations considered in this text in a similar way, and presumably obtain similar results.

See also the extensive work by Popivanov [96–98], concerning equations having the form

$$ K(y)u_{xx} + M(x)u_{yy} + \text{lower-order} = f \tag{B.5} $$

where $u = u(x, y)$, $f = f(x, y)$, $K(0) = M(0) = 0$, $yK(y) > 0$, $xM(x) > 0$, $K'(y) > 0$, and $M'(x) > 0$. The equation is studied from the point of view of symmetric positive differential equations. A multidimensional generalization has also been studied [99].

4. The following two problems concern the auxiliary problem in the integral variant of the *abc* method.

i) What can be said in general about the solution of the Dirichlet problem for first-order hyperbolic systems generalizing (4.44), (4.45)? In addition to its intrinsic interest, this question must be answered if the integral variant of the *abc* method as applied in Sect. 4.3 is to be extended to a large class of equations. The current literature on the Dirichlet problem for hyperbolic equations is inadequate in many respects: First, it tends to be restricted to scalar second-order equations having the form of the wave equation; second, the coefficients must satisfy delicate number-theoretic conditions, whereas the conditions on the coefficients in the system (4.44), (4.45) have to do with the geometry of the domain with respect to a vector field. For a review of the literature on the Dirichlet problem for hyperbolic equations, see [126].

ii) Very recently, M. Khuri introduced an extension of the integral variant of the *abc* method in which the weighted spaces employed in this text, and in the recent literature on this method, are replaced by anisotropic function spaces [50]; Appendix A.6 contains a very brief review of this method. An open-ended direction for future research would be to investigate systematically the effects of this method on the problems reviewed in this text; c.f. the remark at the end of Problem 2.

Anisotropic Sobolev spaces are appropriate for solutions which have greater regularity in one direction than in another. This is typical of elliptic–hyperbolic equations of the form (1.3). Despite its promise, it is unlikely that Khuri's method for exploiting these spaces will entirely replace methods based on the weighted function spaces discussed in earlier chapters. The latter spaces are well-adapted to solutions which have essential singularities on a subset of their domain. Such solutions arise frequently in equations of Keldysh type. For example, solutions to the cold plasma model and to the Laplace–Beltrami equation on mixed Riemannian–Lorentzian metrics are typically singular at one or more points on the sonic curve. Essential singularities also characterize fundamental solutions of Cinquini-Cibrario's equation for certain coefficients of the first-order term.

5. What, in general, is the relation between the vector field $V = -(b, c)$ and the differential equation that permits the integral variant of the *abc* method to work under an assumption that the domain is star-shaped with respect to V? Obviously, this problem is related to Problem 4, but is more fundamental. For example, the choice of multiplier in open boundary value problems for symmetric positive differential equations seems to have a natural interpretation in terms of starlike boundaries; consider, for example, the hypotheses on the boundary in Theorem 3.4.

In [64], invariance of the differential equation under dilations is crucial; see also [91]. But this invariance is not required in applying the method to the closed Dirichlet problem in [58], and is also not exploited in these notes, except in the context of special solutions (Sects. 3.7 and 4.4).

In [58, 85], only vector fields of the form $(b, c) = (mx, \mu y)$ are considered, where m and μ are constants. Can one prove weak existence for more general kinds of elliptic–hyperbolic Dirichlet problems by considering more general classes of vector fields?

The integral variant of the *abc* method was used in [58] to solve the closed Dirichlet problem for a wide class of elliptic–hyperbolic equations of Tricomi type. It was extended in [85] to a particular equation of Keldysh type. An obvious program would be to find the largest class of equations of mixed elliptic–hyperbolic type for which one could show the existence of a weak solution to a closed Dirichlet problem via the integral variant of the *abc* method. Presumably, answering the questions of the preceding paragraphs would contribute to that program. In particular, is there a notion of weak solution that would extend the method discussed in Sect. 4.3 to operators which are not formally self-adjoint?

6. In addition to Problems 4 and 5, the following problems are suggested by the material in Chap. 4:

 i) The nature of the singularity at the origin of the cold plasma model equations is poorly understood. The only results appear to be the analytic results of [95] and the numerical experiments of [77]. Numerical experiments show the presence of singularities propagating away from the origin along characteristic lines. Neither the physical nor the analytic meaning of these singularities is clear at present.

ii) More generally, questions of the dependence of a solution on boundary data are poorly understood for equations of elliptic–hyperbolic type. One expects singularities to be reflected in the boundary conditions that one is able to impose, so an answer to item i) would contribute to understanding how a solution to an elliptic–hyperbolic equation depends on the smoothness of boundary data.

iii) What is the most general type-change function for which Theorem 4.2 remains true for some vector field V? (In particular, see items v) and vi), below.)

iv) All these questions are also interesting for the variant of the model equation having the form

$$\left(x - y^2\right) u_{xx} - u_{yy} + \text{ lower-order terms} = 0, \qquad (\text{B.6})$$

a variant which we generally ignored in this text. The methods of Sect. 4.3 appear to fail utterly in this case.

 v) The heart of the proof of Theorem 4.2 is the fundamental estimate of Lemma 4.1, the proof of which would fail without the term y^2 in the type-change function $\mathscr{K}(x, y) = x - y^2$. Moreover, the singular structure of the solution, which is its most physically and mathematically interesting feature, is apparently derived from the geometry of the parabola $y = x^{1/2}$ at the origin of coordinates. This raises the question of whether there is a larger, mathematically natural class of type-change functions for which these kind of arguments work. For example, what can one say about equations that change type along a conic section? How are the various geometric features of this class of curves reflected in the analysis of the problem? Because the type-change functions of Chap. 5 and 6 are circles, this program would unify the discussions of Chaps. 4–6. Note, however, that different analytic methods are used in Chaps. 4–6.

   Distribution solutions for a closed Dirichlet problem having a type-change function in the form of a hyperbola are studied in the very recent paper [75].

vi) The questions posed in item v) may also be asked about toric sections. That is, can one say something about type-change functions having the general form

$$\mathscr{K}(x, y) = \left(x^2 + y^2\right)^2 + ax^2 + by^2 + cx + dy + e,$$

where a, b, c, d, and e are constants, and if so, how does the geometry of these curves affect the analysis? Of course, such type-change functions are also natural generalizations of those discussed in Chaps. 5 and 6.

vii) Lupo, Morawetz and Payne [58] have established the existence of a unique H^1_{loc} weak solution to the homogeneous closed Dirichlet problem for an inhomogeneous Tricomi equation of the form (1.1). It is known that the corresponding weak solution for the formally self-adjoint form of the cold plasma model (Chap. 4) is not $H^{1,2}$ in any neighborhood of the origin [77]; c.f. item 1 of Sect. A.5. It is natural to ask whether H^1_{loc} solutions to closed boundary value problems exist for *any* equations of Keldysh type having the structure of (1.3).

7. Concerning the discussion of of the fundamental solution for Cinquini-Cibrario's equation in Sect. 3.7:

i) Are there fundamental solutions to the cold plasma model equations? Some of the individual arguments in [9–11, 20, 131] extend to the cold plasma model (Sect. 4.4); but apparent problems arise due to the absence of a closed form for the characteristic lines of the cold plasma model equations.

ii) Fundamental solutions are used to associate an integral equation with a boundary value problem. This integral equation determines a Green's-function solution to the boundary value problem. Thus the existence of a fundamental solution for Cinquini-Cibrario's equation should open the way for the solution of open boundary value problems by the method of Green's functions. (I am grateful to an anonymous referee for making this point.) The boundary condition would be imposed on the elliptic boundary, and appropriate conditions on the domain boundary would be required. The regularity of the fundamental solution described in Sect. 3.7, for various lower-order terms, must be taken into account. The self-adjoint case is particularly interesting, for reasons that are clear from the examples in this text. This problem is, as far as I know, also open for the Tricomi equation, although in that case a more regular fundamental solution exists [9–11], and the Green's function method has already been applied to the degenerately elliptic case [104]. In particular, the authors of [9] suggest that unnatural restrictions on the boundary that accompany the classic papers on the existence of solutions to open boundary value problems for Tricomi's equation could be removed by the use of the Barros-Neto–Gelfand solution; a similar hope exists for Green's functions created by integrating the Chen solution to Cinquini-Cibrario's equation over a suitable domain.

8. Recall that we use the term *elliptic boundary* to refer to that part of the domain boundary on which the type-change function is positive. Similarly, by the *hyperbolic boundary* we mean the collection of boundary arcs for which the type-change function is negative. We define a *fully elliptic–hyperbolic* boundary value problem to be a boundary value problem for which the hyperbolic boundary is non-empty and the subset of the hyperbolic boundary on which data have been prescribed is

also non-empty. Note that this definition is independent of whether the boundary value problem is open or closed.

The fact that the data in Theorems 5.1 and 5.2 are prescribed only on the elliptic boundary is an important restriction, as it is the hyperbolic boundary on which the Dirichlet problem typically becomes over-determined. So there is obvious interest in extending Theorems 5.1 and 5.2 to fully elliptic–hyperbolic boundary value problems. For example, does a solution to the equations of Theorems 5.1 and 5.2 exist in an annulus about the sonic curve, as was shown for the Tricomi-type equation in [121]? One would also like a result for harmonic fields on the extended projective disc which is analogous to Theorem 5.1. In that case the boundary would be fully hyperbolic, which is much harder, but again one could look for a solution in an annulus.

Also regarding harmonic fields on the extended projective disc, there should be a proof of the existence of weak solutions to an open Dirichlet problem which does not require a rearrangement of the lower-order terms between the two equations in the system (c.f. Sect. 6.3.1).

9. There is no obvious reason why one should not have maximum principles for equations of Keldysh type which are analogous to those derived in [3, 62] for equations of Tricomi type; but there do not seem to be any such results in the literature. (I am indebted to Yuxi Zheng for this observation.) Similarly, eigenvalue problems for the equations of Sect. 6 are only known for very special cases [130] and appear to be completely unknown for the equations of Chaps. 2–4; c.f. Problem 12.

10. The existence and nature of water caustics have been the subject of both theoretical and experimental research. In Sect. 3.3.1 of [73], trapped water waves on a ridge and in a submarine trough are analyzed. It is shown that the geometrical optics approximation is valid and that the rays form an envelope, producing a caustic; see also [8, 105]. These analyses have been supported by experimental studies of ocean caustics [18, 93]. To what extent do the methods of Chap. 5 extend to water caustics?

There are certain apparent obstructions. For example, the refractive index is highly variable in realistic applications to water waves (see below). Also, water caustics often occur in the context of shoaling waves, in which turbulent effects tend to work against optical analogies. Nevertheless, an apparently reasonable model based on a Ludwig–Kravtsov-like system has been formulated by Chao for water waves near a caustic [17] – see also [18]. Chao's model proceeds from the fact that the equations of motion for an inviscid and incompressible liquid in simple harmonic motion, bounded below by an impervious and rigid bottom, can be written in the linearized, dimensionless form [74]

$$\lambda^2 \varphi_{zz} + \Delta \varphi = 0, \quad -h(x, y) \leq z \leq 0. \tag{B.7}$$

At $z = -h(x, y)$,

$$\lambda^2 \varphi_z + \nabla h \cdot \nabla \varphi = 0, \tag{B.8}$$

and at $z = 0$,

$$\varphi_z = \varphi \tag{B.9}$$

and

$$\tilde{\eta} = \mathrm{Re}\left(-i\varphi e^{-i\tau}\right). \tag{B.10}$$

Here $\lambda = \mathscr{L}\omega^2/g$, where ω is the angular frequency of the wave, g is the acceleration due to gravity, and $\mathscr{L}$ is the horizontal scale length of the bottom contours; $\varphi = \nu\Phi$, where $\Phi = \Phi(x, y, z)$ is the velocity potential of the wave in cartesian coordinates (x, y, z), z is the vertical axis positive upward from the equilibrium water level, and $\nu = \omega^3/g^2$; h is the product of the water depth and the scaling factor $\lambda/\mathscr{L}$; $\tilde{\eta}$ is the product of the surface fluctuation and the same scaling factor; $i^2 = -1$; and $\tau = \omega t$, where t is time.

Expressing φ in the form

$$\varphi = w(x, y) \cosh\left[k(h + z)\right], \tag{B.11}$$

where k is a re-scaling of the wave number by the quantity $\lambda\mathscr{L}$, we notice that $\varphi_{zz} = k^2\varphi$ and (B.7)–(B.10) imply (5.12) in the form

$$\left(\Delta + k^2\lambda^2\right)\varphi = 0.$$

In this case the geometrical optics approximation is equivalent to letting λ tend to infinity. By the definition of λ, this would require the horizontal scale length of the bottom to be large relative to the deep-water wavelength $2\pi g/\omega^2$. Moreover, the arguments of [49], Sect. 3, can be used to obtain the identity

$$\lambda = (kh/S_b) \tanh kh,$$

where S_b denotes the bottom slope. This implies that, for fixed kh, λ will be large when the bottom is close to horizontal. Based on these considerations, Chao [17] estimates that that the geometrical optics approximation is reasonable for most areas of shoaling water except those in the neighborhood of the shoreline, in which regions the linearization itself fails as a result of surf effects. This form of the Helmholtz equation is not only mathematically equivalent to the equation satisfied by standing waves in wave optics; the physical interpretations of the coefficients k, λ are analogous as well, in the sense that the caustic associated to a shoaling wave is a refractive effect.

Condition (B.11) fails at a caustic, motivating the Ludwig–Kravtsov ansatz. In this context the Kravtsov–Ludwig ansatz consists in expressing the leading term of the velocity potential φ, for large λ, in the form

$$\varphi(x, y) = e^{i\lambda\theta(x,y)} \cosh\left[k(h + z)\right]$$
$$\times \left\{ \gamma_0(x, y)A\left[\lambda^{2/3}\rho(x, y)\right] + \frac{i\,\gamma_1(x, y)}{\lambda^{1/3}}A'\left[\lambda^{2/3}\rho(x, y)\right] \right\},$$

where the notation ρ, θ, γ_0, γ_1, and A is as in Sect. 5.1. We obtain from the Helmholtz equation an expression of the form

$$(\nabla\theta)^2 - \rho\,(\nabla\rho)^2 - k^2 = \nabla\theta \cdot \nabla\rho = 0. \tag{B.12}$$

Because we assume that the propagation speed is approximately equal to a nonzero constant c over a sufficiently short interval of time, we can replace the wave number k by a constant inversely proportional to c. Under such an assumption it is convenient to re-scale θ and ρ by defining new variables $\tilde{\theta} = k^{-1}\theta$ and $\tilde{\rho} = k^{-2/3}\rho$. In that case we can divide (B.12) by k^2 and replace the term k by the number 1. This results in an equation identical to (5.42).

Thus it is natural to wonder about the extent to which the arguments of [24, 25, 66–70] are applicable to Chao's model. A related question is the extent, if any, to which the method of evanescent wave tracking and complex ray tracing (see, e.g., [31, 32]) is applicable to water waves.

A potential difficulty in the water wave case is the inversion of the solution in the hodograph plane. It is likely that there will be some nonlinear terms in the physical plane which must be neglected if the system is to be put into the homogeneous form required for the hodograph transformation. In the case of optics, the refractive index can be taken to be constant, which forces the inhomogeneous terms in the nonlinear equation to vanish. That simplification does not seem to be realistic in the case of water waves.

We note that water caustics are not only caused by the propagation of deep-water waves into shoaling water. For example, circular water caustics may arise in the contexts of a marine explosion or the impact of a high-velocity body; c.f. [13, 127].

11. The distance element associated to the metric for a stationary, rotating, axisymmetric field has the general form [56]

$$ds^2 = -A(r)dt^2 + 2B(r)d\varphi dt + C(r)d\varphi^2 + D(r)\left(dr^2 + dz^2\right). \tag{B.13}$$

This metric is Lorentzian provided the matrix determinant

$$g = -\left(AC + B^2\right)D^2$$

is negative.

Examples include the *van Stockum metric* [118], for which $A(r) = 1$, $B(r) = \omega r^2$, $C(r) = r^2\left(1 - \omega^2 r^2\right)$, and $D(r) = \exp[-\omega^2 r^2]$. This metric arises in a model of a rotating infinite cylinder of dust, held stationary within a vacuum by the balance of centrifugal and gravitational forces on the dust. Because the variable φ in (B.13) is the angular coordinate, any azimuthal curve for which the variables t, r, and z are held constant will have invariant length

$$ds^2 = 4\pi^2 C(r).$$

The resulting integral curve will be a closed, time-like curve provided $C(r) < 0$. The role of such curves in van Stockum's model was apparently first realized by Tipler [119], long after the model was originally introduced. Closed, time-like curves are associated with certain violations of causality and for that reason have attracted interest.

The z-coordinate, representing the axis of rotation, is not very important in the analysis of the metric (B.13). If we ignore this variable, then in the stationary case we obtain, for $D(r) = 1$, the distance formula for the extended projected disc:

$$ds^2 = r^2 \left(1 - \omega^2 r^2\right) d\varphi^2 + dr^2.$$

In this interpretation the angular velocity in van Stockum's model has become the hyperbolic curvature $K = -\omega^2$.

Writing (6.16) in its second-order form, we obtain the wave equation on the curved metric g (*i.e.*, the Laplace–Beltrami equation)

$$\frac{1}{\sqrt{|g|}} \frac{\partial}{\partial x^i} \left(g^{ij} \sqrt{|g|} \frac{\partial u}{\partial x^j} \right) = 0. \tag{B.14}$$

This equation is often reduced by imposing an invariance condition which amounts to a consideration of stationary waves; see, e.g., Sect. 2 of [121]. Applying this operator to the metric associated to (B.13), we obtain the differential operator of (1.4) with $\mathscr{K}(r) = C/D$ and $k = 1/2$.

Thus in particular, we know from Theorem 3.1 not to expect classical solutions to a closed Dirichlet problem on a domain such as the one constructed in Sect. 3.3. It is therefore natural to ask whether strong solutions exist, by applying the approach of Theorem 5.2.

Applying the first-order form of the equation, (6.16), with $u_1 = u_x$, $u_2 = u_y$, and taking $A(t) = B(t) = 0$ in (B.13) we obtain, ignoring the z-coordinate, the system

$$\left[\frac{(|C|D)_r - 2|C|D_r}{2|C|D^2} \right] u_1 + \frac{u_{1r}}{D} + \frac{u_{2\varphi}}{C} = f \tag{B.15}$$

$$u_{1\varphi} - u_{2r} = 0. \tag{B.16}$$

In order to apply the methods of Theorem 5.2, we might consider this system in the disc

$$\Omega = \{(r, \varphi) \,|\, 0 \le r \le R, 0 < \varphi \le 2\pi\},$$

and assume that $D(r)$ exceeds zero on Ω and that $C(r)$ changes from positive to negative sign on a circle $r = r_{crit}$ in the interior of Ω. The given function f on the right-hand side of (B.15) has been inserted for mathematical generality; see also Sect. 2.6. Multiplying (B.15) by C and carrying out the indicated operations, we obtain the simpler equation

$$Mu_1 + \frac{C}{D}u_{1r} + u_{2\varphi} = \tilde{f}, \tag{B.17}$$

where

$$M = \frac{C_r D - CD_r}{2D^2}$$

and $\tilde{f} = Cf$. It is easy to check that this system is not symmetric positive and that the methods used to prove Theorem 5.2 do not apply in any obvious way.

Similarly, the methods used in Chap. 4 to obtain weak solutions do not have an obvious application. Thus it remains an interesting open problem to determine conditions under which weak or strong solutions exist for the Laplace–Beltrami equations on a relativistically rotating disc on which closed, time-like lines are permitted.

12. Lupo, Payne, and occasional collaborators have studied the spectral properties of linear and semilinear Tricomi operators [57, 59, 63]. I know of no such results for elliptic–hyperbolic operators of Keldysh type.

13. An existence theorem for the closed Neumann problem has been proven for the Lavrent'ev–Bitsadze equation (2.5) by Pilant [94]. No other result for conormal conditions on the entire boundary is known for elliptic–hyperbolic equations of either type.

B.1 Nonlinear Equations

The problem of extending these methods to nonlinear equations deserves a discussion of its own.

14. *A nonlinear extension of problem 10:* The focusing of wave action in a caustic region has been advanced as an explanation for giant *rogue waves* which have been observed, for example, in the Agulhas current off the southeast coast of Africa; see [36, 53, 92, 108, 128]. In this case as well, related experimental studies have been conducted; see, e.g., [38] and, for deep-water examples, the statistical studies cited in Sect. 1 of [27].

In studying the interaction of a wave with an opposing current, a ray approximation is indicated by comparison of the length of even large wind-induced swells, such as the ones correlated with accidents in the Agulhas, with the scale of horizontal variations of the current. (This point is made in [108], p. 417.) Models of rogue waves based on the theory of caustics have been criticized on different grounds: that the incoming ocean wave would have to enter the zone of variable currents with a single direction. Otherwise, the wave would be too diffuse to focus into a caustic; c.f. [27], Sect. 3.

An analogy can be drawn to optical caustics at the bottom of a swimming pool on a *sunny* day (unidirectional incoming wave) versus their absence on a *cloudy* day (diffuse incoming wave). In a swimming-pool caustic, the light waves scatter by refraction through the water. In the model of rogue waves, the incoming wave is refracted by collision with the fast-moving Agulhas current. A sunny day would correspond to a unidirectional current, which would presumably be very rare (maybe too rare). A cloudy day would correspond to the usual case of a current composed of individual waves moving in multiple directions. This analysis suggests that focusing *can* produce extreme waves; what fraction of the observed examples actually *is* produced by them is not so clear. As a reaction to this kind of criticism, recent papers have concentrated on a statistical analysis of the probability of a caustic-producing current [37]; see also the recent experimental work reported in [120].

However, this problem may also be attacked through a better understanding of the relevant partial differential equations, which remain poorly understood. There exist deep-water models for effects exerted on water waves by reflection of rays at a caustic, which take into account diffractive effects arising from nonlinear terms in the governing equations. Those models are based on the observation that the governing equations for the wave amplitude can be put into the approximate form of a nonlinear Schrödinger equation [132]. Using this kind of approach, it has been estimated that a three-fold amplification of wave amplitudes could occur near a caustic in the collision of a wave with the Agulhas Current [108]. Moreover, rogue waves in this model possess an asymmetry that accounts for reports of a deep trough that precedes the steep forward face observed in rogue waves. The presence of such a trough, and the asymmetrical steepness of the forward face of the wave, explain some of the destructive effects of such waves on oil tankers. (No change of equation type is associated with such models.)

But extreme waves also occur in shallower waters. So it is possible that a better understanding of shallow-water models such as the one described in item *x*) of Sect. 2.7.2 might also be welcome. These have been used in studying, for example, the focusing of tidal waters by a narrowing at the mouth of channel, inducing a change in their velocity as the tide enters the still water of the channel [123]. A similar effect occurs as a wave progresses up a sloping beach. In each case the change in the velocity profile corresponds to the focusing effects in optics which arise from changes in the refractive index of the medium. A geometrical optics approximation is reasonable in this case as well, as the relevant dimension for applying the geometrical optics approximation is the ratio of the width of the channel entrance to the horizontal scale of the incoming tide. In cases for which this ratio is small, ray effects will dominate over diffractive effects at the mouth of the channel. Some technical issues regarding the geometrical optics approach in shallow water theory are addressed in Sect. 1 of [49].

While there is already a large interdisciplinary literature on various aspects of extreme waves, the literature on elliptic–hyperbolic transitions in hydrodynamics is considerably smaller. A possible resource is the large literature on elliptic–hyperbolic transitions and shock waves in gas dynamics, which are analogous in some respects, but not analogous in others.

15. Existence/nonexistence theorems for semilinear forms of Cinquini-Cibrario's equation, even in the degenerate elliptic form studied by Keldysh, would presumably have applications to the study of magnetically dominated plasmas. (See, e.g., the discussion following (16), (17) in [125]; there is interest in various contexts for $\overline{Q}$ having a polynomial dependence on $\overline{P}$ in (16) of that reference. A similar remark applies to the discussion preceding (8) in [40].) An approach to the nonexistence question might be to try to extend the methods of [64, 65] to certain equations of Keldysh type. Tricomi problems for semilinear equations of Tricomi type are investigated in [60]; those results may have extensions to equations of Keldysh type.

16. A quasilinear theory of symmetric positive operators is outlined in [34], the Appendix to [35], and [124]; but the quasilinear case has been much less intensively studied than the linear case, both in terms of its mathematical properties and its applications. Quasilinear elliptic–hyperbolic problems in general, other than those associated with gas dynamics, have not been studied very much except in linearized forms. Recent exceptions include [22, 23]. Applications of methods originating in gas dynamics to various problems in general relativity have been pioneered by Smoller in papers with various collaborators; see, e.g., [110]. The hypotheses in those papers are, appropriately, intimately connected with the physical model, and it is not clear how they would apply to a general mathematical theory for equations of Keldysh type. Of course the distinction between Keldysh type and Tricomi type applies to linear and, by an obvious extension, semilinear equations. However, certain quasilinear equations become equations of Keldysh type in a natural linearization. An example (in addition to those in Chaps. 5 and 6) is

$$\left((u + x)\, u_x - \frac{u}{2} \right)_x + u_{yy} = 0$$

for $u = u(x, y)$, which is studied in [15, 16].

We note that recently, Friedrichs' criteria for admissible boundary conditions have been related to three apparently different sets of intrinsic geometric conditions in graph spaces [30]. Subsequently, those intrinsic boundary conditions were shown to be equivalent when reinterpreted in the context of Kreĭn spaces [5]. These papers are examples of a resurgence of interest in symmetric positive operators among theoretical numerical analysts; see also [14,29,41]. It is possible that these reformulations of the theory in more abstract contexts will facilitate quasilinear extensions.

17. The technique reviewed in Sect. 2.6, for deducing the existence of solutions to an inhomogeneous Dirichlet problem for a homogeneous equation by solving a homogeneous Dirichlet problem for the corresponding inhomogeneous equation, does not extend to nonlinear equations. Thus the entire issue of Dirichlet problems having inhomogeneous data becomes important for nonlinear equations to a degree that is not encountered in the linear case. As noted in Sect. 5.3, inhomogeneous boundary conditions become more complicated under linearization of the associated equation by the hodograph map; so difficulties in boundary conditions for nonlinear equations are only rarely improved by a hodograph linearization, even under the restrictive conditions in which such linearizations are possible.

18. Because the ellipticity condition for Euler–Lagrange equations can double as a convexity condition for many of the standard energy functionals, variational approaches tend to be associated with equations of elliptic type. But, for example, General Relativity is a variational theory having strictly hyperbolic variational equations; and, as illustrated in Sect. 2.7, a large class of quasilinear elliptic–hyperbolic equations can be derived as Euler–Lagrange equations of an energy functional. We will consider two candidates for further research – one involving semilinear equations and the other involving quasilinear equations:

i) Variational methods for open boundary value problems involving semilinear equations of Tricomi type have been pioneered by Lupo and Payne ([60]; see also [57, 61]). The papers employ a *dual variational method*, in which a semilinear partial differential equation of the form $Lu = F(u)$, having variational structure, is solved by first inverting the corresponding linear operator and then treating the resulting equation

$$u = L^{-1}F(u) \tag{B.18}$$

by variational means.

Consider an open Tricomi boundary value problem of the form considered in (3.1)–(3.3), but with a nonlinear term $Rf(u)$ added to (3.1). Here R is the reflection operator on $L^2(\Omega)$ induced by composition with the map taking points (x, y) of $\mathbf{R}^2$ into points $(-x, y)$, where Ω has the general form of the domain in Fig. 3.1. In particular, Ω is symmetric about the y-axis and is bounded smoothly by a Jordan curve in the elliptic region of the equation and piecewise smoothly by two characteristic lines Γ_1 and Γ_2.

For a fixed arc $\Gamma \in \partial\Omega$, the associated Sobolev space is the closure of H^1_Γ with respect to the H^1-norm of the set $C_\Gamma(\Omega)$ consisting of smooth functions on $\overline{\Omega}$ vanishing identically on Γ. This space has a dual, H^{-1}_Γ, formed in the expected way (c.f. [61], (1.1)).

The fundamental difficulty with applying variational methods in this context is that the Tricomi operator for an open boundary value problem does not map H^1_Γ onto its dual, but rather onto the dual of the adjoint problem. If, for example, the solution is constrained to vanish on the elliptic boundary and a characteristic curve Γ_1, then the dual of the adjoint problem has data vanishing on the elliptic boundary and the characteristic curve Γ_2. In the dual variational method, the Tricomi operator is composed with the operator R, which induces an isometric isomorphism between the adjoint boundary spaces. In this way, a variational structure can be associated with the boundary value problem. The underlying idea is to then take advantage of the compactness of the inverse of the linear Tricomi operator T by solving (B.18) for $L = RT$ and $F = Rf$.

One question is whether this method can be extended to treat equations associated with different or more general elliptic–hyperbolic operators. But there are also interesting avenues of research for the dual variational method in its current application to the Tricomi operator. For example, because it is necessary to obtain the continuity of the Nemitskii operator associated to the nonlinearity, this method requires the nonlinearities to have at most asymptotically linear growth. In order to treat nonlinearities having superlinear growth, one would have to develop a suitable L^p theory for the linear Tricomi operator; see [61], Sect. 1.

ii) The nonlinear Hodge–Frobenius equations (5.76), (5.77) for a differential form of arbitrary degree can be derived by applying variational arguments to a nonlinear Hodge energy functional as in Sect. 2.7.1. We obtain ([72], Sect. 5.1.1)

$$d * [\rho(Q)\omega] = -d\eta \wedge * [\rho(Q)\omega], \tag{B.19}$$

where now ω is a gradient-recursive k-form and η is a 0-form. The other notation is as in Sect. 5.6.1. This is a quasilinear equation of roughly similar form to the class considered by Tso [124]. It is natural to wonder whether this quasilinear variational equation, which for appropriate ρ may be of elliptic–hyperbolic type, can be treated by the quasilinear extensions of the theory of symmetric positive operators introduced by Gu and Tso.

A condition broadly analogous to the Frobenius condition (5.77) arises if ω is taken to be a Lie-algebra-valued 2-form F_A, where A is a Lie-algebra-valued 1-form. In that case, the second Bianchi identity

$$dF_A = -[A, F_A], \tag{B.20}$$

where $[,]$ denotes the Lie bracket, has a form analogous to (5.77).

Precisely, let X be a vector bundle over a smooth, finite, oriented, n-dimensional Riemannian manifold M. Suppose that X has compact structure group $G \subset SO(m)$. Let $A \in \Gamma(M, ad\, X \otimes T^*M)$ be a connection 1-form on X having   curvature 2-form

$$F_A = dA + \frac{1}{2}[A, A] = dA + A \wedge A,$$

where $[,]$ is the bracket of the Lie algebra $\mathfrak{I}$, the fiber of the adjoint bundle $ad\, X$. Sections of the automorphism bundle $Aut\, X$ are *gauge transformations,* acting tensorially on F_A but affinely on A; see, *e.g.,* [71].

One can form energy functionals analogous to the nonlinear Hodge energy, in which $Q = |F_A|^2 = \langle F_A, F_A \rangle$ is an inner product on the fibers of the bundle $ad\, X \otimes \Lambda^2(T^*M)$. The inner product on $ad\, X$ is induced by the normalized trace inner product on $SO(m)$ and that on $\Lambda^2(T^*M)$, by the exterior product $*(F_A \wedge *F_A)$.

A nonabelian variational problem analogous to (5.77), (5.76) is described briefly in Sect. 5.1 of [84]. One is led to consider smooth variations having the form

$$var\,(E) = \int_M \rho(Q) var(Q) dM = \int_M \rho(Q) \frac{d}{dt}_{|t=0} |F_{A+t\psi}|^2 dM$$

$$= \int_M \rho(Q) \frac{d}{dt}_{|t=0} |F_A + tD_A\psi + t^2\psi \wedge \psi|^2 dM,$$

where $D_A = d + [A,\ ]$ is the exterior covariant derivative in the bundle. The Euler–Lagrange equations are

$$\delta\,(\rho(Q)F_A) = -*[A, *\rho(Q)F_A]. \tag{B.21}$$

In addition, we have the Bianchi identity (B.20).

If we write (5.77) in components

$$d\omega^a = \Gamma^a_b \wedge \omega^a,$$

then if $-\Gamma$ is interpreted as a connection 1-form, (5.77) can be interpreted as the vanishing of an exterior covariant derivative, which is the analytic content of (B.20). Moreover, we recover the well known algebraic requirement that Γ must satisfy

$$\left(d\Gamma^a_b - \Gamma^a_c \wedge \Gamma^c_b\right) \wedge \omega^b = 0$$

(c.f. (4-2.3) of [28]) as a zero-curvature condition:

$$[F_\Gamma, \omega] = 0.$$

If a suitable modification of the methods of [124] can be applied to (B.19), (5.77), then the question arises whether they can be applied to the nonabelian extension (B.20), (B.21). It may be necessary to introduce a suitably large lower-order perturbation in order for there to be any hope of using these methods.

On the one hand it seems absurd to suggest the study of the elliptic–hyperbolic form of systems such as (B.20), (B.21) in a text which devotes an entire chapter to an equation as simple as the cold plasma model equation (4.35). But on the other hand, the equations of motion for a plasma in their full generality are not noticeably simpler than the system (B.20), (B.21). A major feature of the study of such a system is the search for physically or geometrically reasonable special cases in which the equations simplify to analytically tractable forms.

19. The prescription of asymptotic boundary conditions at infinity is the global analogue of the Dirichlet problem. Variational conditions which lead to analytic results under growth hypotheses at infinity are notable for being at least superficially independent of type, especially if the variations are taken by reparametrization of the underlying domain – so-called *r-variations* [4] – rather than in the infinitesimal deformation space of the solution.

For example, consider an energy integral having the very general form

$$E = \int_M w\left(|du|^p\right) dv_g, \quad p > 0,$$

where M is a Riemannian manifold having local metric g and u is a map from M into another Riemannian manifold N. This functional was introduced in [80] in the context of a Liouville theorem for a stationary point under r-variations, with prescribed growth conditions on the energy. The function w is assumed to satisfy the conditions

$$0 \leq \dot{w}(t) \leq K_1, \tag{B.22}$$

$$K_2 t \leq w(t) \tag{B.23}$$

for constants $K_1 \geq 0$ and $K_2 > 0$. The map u is an example of what would later be called an F-*harmonic map* [6]. Condition (B.23) and the boundedness of the derivative in condition (B.22) are not a feature of F-harmonic maps in general; they are imposed in order to derive the Liouville theorem, which is not satisfied for an r-stationary F-harmonic map without some extra hypotheses. The type of the variational equations of this object is not strictly specified without some further condition on the function w.

In the special case in which

$$w\left(|du|^p\right) = \int_0^Q \rho(s)ds,$$

where $Q = |du|^2$, the functional E reduces to the nonlinear Hodge energy of Sect. 2.7.1 (modulo the multiplicative constant 1/2). However, the variations satisfied by r-stationary points of E are quite different from those which produce the nonlinear Hodge equations (5.71), (5.72), as the latter are taken in the infinitesimal deformation space of the map rather than in the reparametrization space of the underlying domain.

Asymptotic conditions are imposed on both the manifold and the energy in [80] under which r-stationary points are forced to be globally trivial.

One purpose of Liouville theorems is to decide how singular a geometric object must be in order to avoid global triviality. For this reason, the domain manifold in [80] is allowed to have a singular set of prescribed Hausdorff dimension. In addition to mappings, Liouville theorems have been established for variational points that live on a vector bundle, and which are associated, by a particular choice of mass density, to energies of generalized Yang-Mills or Yang-Mills–Born-Infeld type. See, e.g., Sect. 5 of [84] and the references therein – also Sect. 1 of [83] and Sect. 4 of [107]. The very recent paper [26] introduces an associated triviality condition for the Dirichlet problem and a collection of vanishing theorems for differential forms. See also Sect. 2 of [82], Sect. 1 of [39], and Sect. 4.2 of [72].

It is natural to ask what geometric and analytic hypotheses would be necessary and/or sufficient in order to extend these results from the case of Riemannian manifolds to a class of semi-Riemannian manifolds. Another obvious application of the results would be to apply them to specific densities of physical or geometric interest, such as those cited in Sect. 2.7.

20. I am indebted to Yisong Yang for drawing my attention to the following problem, which arises in quantum field theory [1, 2].

By a series of fortunate approximations, the partition function for quantum electrodynamics can be reduced to a relativistic model in which the quarks are coupled to a pair of classical gauge fields. The latter are represented by a vector-valued gauge potential **A** and a scalar-valued field ϕ. (Note that a "scalar" in this context is what mathematicians might call a *weighted relative tensor* – see, e.g., the discussion on pp. 105–107 of [81].) Additional choices reduce the analysis to a problem in nonlinear electrostatics with the governing equations

$$\nabla \cdot (\sigma \nabla \Phi) = 0, \tag{B.24}$$

where Φ is a – suitably interpreted – flux, and σ a prescribed function of the cylindrical radial coordinate ρ and the scalar $|\nabla \Phi|$. This equation is superficially similar to (2.35) of Sect. 2.7; but the function

$$\sigma = \sigma\,(\rho, |\nabla \Phi|)$$

is *not* assumed to be a quadratic function of $\nabla \Phi$; see (20b) and (21) of [2].

Defining the inward-pointing unit normal and its normal derivative in terms of Φ via

$$\hat{\mathbf{n}} = \frac{\nabla \Phi}{|\nabla \Phi|}, \quad \partial_n = \hat{\mathbf{n}} \cdot \nabla,$$

it can be shown that (B.24) reduces to

$$\left[\partial_\rho^2 + \partial_z^2 + (\alpha - 1)\,\partial_n^2 \right] \Phi - \alpha \rho^{-1} \partial_\rho \Phi = 0, \tag{B.25}$$

where

$$\alpha = 1 + \frac{\partial \log \sigma}{\partial \log |\nabla \Phi|}.$$

Defining $\hat{\ell}$ to be the unit tangent vector to the surface of constant Φ and ∂_ℓ to be the corresponding tangential derivative, the differential operator of (B.25) can be put into the form

$$L\Phi = \left[\partial_\ell^2 + \alpha \partial_n^2 \right] \Phi.$$

That is, L is elliptic, hyperbolic, or parabolic, depending on whether α is positive, negative, or zero. The nonlinearities of this operator are considerably wilder than those of the operators studied in this text, and it is not clear how L can be effectively linearized. However, the model underlying this analysis is very rich, and opportunities may be found for further simplification.

For example, in the *leading logarithm model* considered in [1, 2], still more choices reduce (B.25) to a degenerately elliptic equation. It is not clear whether (B.25), which applies to any effective action density depending quadratically on the electric field, is accessible without imposing severe physical restrictions on the generality of the model.

21. The final problem is, in a sense, the most difficult. It is the problem that the linear theory of elliptic–hyperbolic equations yields relatively little qualitative insight into the nonlinear theory. Physical models of classical fields are generally nonlinear, which means that they are in most cases beyond the reach of linear analysis.

This is similar to the case of hyperbolic differential equations. But it is in sharp distinction to the theory of elliptic (and parabolic) differential equations, in which the qualitative behavior of the potential equation (and heat equation) is a rough guide to the corresponding nonlinear theory; c.f. [33, 52, 78].

Consider for example the nonlinear equations which arise in the two-dimensional, compressible Euler equations for an ideal fluid. We briefly review a few aspects of this vast and complex topic, following [133, 135]. The governing equations in this case are:

i) The conservation of mass:

$$\rho_t + (\rho u)_x + (\rho v)_y = 0;$$

ii) The conservation of linear momentum in u :

$$(\rho u)_t + \left(\rho u^2 + p\right)_x + (\rho u v)_y = 0;$$

iii) The conservation of linear momentum in v :

$$(\rho v)_t + (\rho u v)_x + \left(\rho v^2 + p\right)_y = 0;$$

iv) The conservation of energy:

$$(\rho E)_t + (\rho u E + u p)_x + (\rho v E + v p)_y = 0,$$

where

$$E = \frac{u^2 + v^2}{2} + e,$$

for $(x, y) \in \mathbf{R}^+$, $t \in \mathbf{R}$. Here e denotes the internal energy of the system. For a polytropic gas,

$$e = \frac{p}{(\gamma - 1)\rho},$$

where $\gamma > 1$ is the adiabatic constant of the gas.

A simpler model, in which we take $\rho = 1$ and ignore inertial contributions, is the *pressure-gradient system*:

$$u_t + p_x = 0,$$
$$v_t + p_y = 0,$$
$$E_t + (pu)_x + (pv)_y = 0.$$

Assuming that solutions are sufficiently smooth and applying the coordinate transformation

$$p = (\gamma - 1)\, P, \ t = \frac{T}{\gamma - 1},$$

this system reduces to the single equation

$$\left(\frac{P_T}{P}\right)_T - (P_{xx} + P_{yy}) = 0. \tag{B.26}$$

The Cauchy problem for even this relatively simple form of the equations is an open problem.

There is also a hydrodynamic interpretation of this model, representing flow in a shallow channel with a bump [21]; c.f. Problem 14.

Some progress has been made in the analysis of self-similar solutions, which occur naturally in many contexts. For example, it is often useful to seek solutions depending on the new variables

$$\xi = \frac{x}{T}, \ \eta = \frac{y}{T}$$

in which case (B.26) assumes the form

$$\left(P - \xi^2\right) P_{\xi\xi} - 2\xi\eta P_{\xi\eta} + \left(P - \eta^2\right) P_{\eta\eta} + \frac{\left(\xi P_\xi + \eta P_\eta\right)^2}{P} - 2\left(\xi P_\xi + \eta P_\eta\right) = 0. \tag{B.27}$$

Near the origin this equation is elliptic, but far from the origin, it is hyperbolic. The well-posedness of boundary value problems outside of the elliptic regime is a rich source of open problems. Another source of open problems is the reflection of the wave off of obstacles having various geometries; see, e.g., [19, 134] and references therein.

Consideration of such problems, even on a superficial level, would take us very far afield, and for that reason it is natural to end our course here.

References

1. Adler, S.L., Piran, T.: Relaxation methods for gauge field equilibrium equations. Rev. Mod. Phys. **56**, 1–40 (1984)
2. Adler, S.L., Piran, T.: Flux confinement in the leading logarithm model. Phys. Lett. **113B**, 405–410 (1982)

3. Agmon, S., Nirenberg, L., Protter, M.H.: A maximum principle for a class of hyperbolic equations and applications to equations of mixed elliptic–hyperbolic type. Commun. Pure Appl. Math. **6**, 455–470 (1953)
4. Allard, W.K.: On the first variation of a varifold. Ann. of Math. **95**, 417–491 (1972)
5. Antonić, N., Burazin, K.: Intrinsic boundary conditions for Friedrichs systems. Commun. Part. Differ. Equat. **35**, 1690–1715 (2010)
6. Ara, M.: Geometry of F-harmonic maps. Kodai Math. J. **22**, 243–263 (1999)
7. Arnold, V.I., Khesin, B.A.: Topological Methods in Hydrodynamics. Springer, Berlin (1998)
8. Arthur, R.S.: Refraction of water waves by islands and shoals with circular bottom contours. Trans. Amer. Geophys. Union **27**, 168–177 (1946)
9. Barros-Neto, J., Gelfand, I.M.: Fundamental solutions for the Tricomi operator. Duke Math. J. **98**, 465–483 (1999)
10. Barros-Neto, J., Gelfand, I.M.: Fundamental solutions for the Tricomi operator II. Duke Math. J. **111**, 561–584 (2002)
11. Barros-Neto, J., Gelfand, I.M.: Fundamental solutions for the Tricomi operator III, Duke Math. J. **128**, 119–140 (2005)
12. Bateman, H.: Notes on a differential equation which occurs in the two–dimensional motion of a compressible fluid and the associated variational problems. Proc. R. Soc. London Ser. A **125**, 598–618 (1929)
13. Berry, M.V.: Tsunami asymptotics. New J. Phys. **7**, 129–147 (2005)
14. Burazin, K.: Contribution to the theory of Friedrichs' and hyperbolic systems, Ph.D. thesis, University of Zagreb (2008)
15. Čanić, S., Keyfitz, B.: A smooth solution for a Keldysh type equation. Commun. Part. Diff. Equat. **21**, 319–340 (1996)
16. Čanić, S., Keyfitz, B.: An elliptic problem arising from the unsteady transonic small disturbance equation J. Diff. Equations **125**, 548–574 (1996)
17. Chao, Y-Y.: An asymptotic evaluation of the wave field near a smooth caustic. J. Geophys. Res. **76**, 7401–7408 (1971)
18. Chao, Y-Y., Pierson, W.J.: Experimental studies of the refraction of uniform wave trains and transient wave groups near a straight caustic. J. Geophys. Res. **77**, 4545–4554 (1972)
19. Chen, G-Q., Feldman, M.: Global solutions of shock reflection by large-angle wedges for potential flow. Ann. of Math. **171**, 1067–1182 (2010)
20. Chen, S-X.: The fundamental solution of the Keldysh type operator, Science in China, Ser. A: Mathematics **52**, 1829–1843 (2009)
21. Cole, J.D., Cook, L.P.: Transonic Aerodynamics. North-Holland, Amsterdam (1986)
22. Dechevski [Dechevsky], L.T., Popivanov, N.: Morawetz-Protter 3-D problem for quasilinear equations of elliptic-hyperbolic type. Critical and supercritical cases. C. R. Acad. Bulgare Sci. **61**, 1501–1508 (2008)
23. Dechevski [Dechevsky], L.T., Popivanov, N.: Quasilinear equations of elliptic-hyperbolic type. Critical 2D case for nontrivial solutions. C. R. Acad. Bulgare Sci. **61**, 1385–1392 (2008)
24. De Micheli, E., Vianoz, G.A.: The evanescent waves in geometrical optics and the mixed hyperbolic-elliptic type systems. Applicable Anal. **85**, 181–204 (2006)
25. De Micheli, E., Vianoz, G.A.: Geometrical theory of diffracted rays, orbiting and complex rays. Russian J. Mat. Phys. **13**, 253–277 (2006)
26. Dong, Y., Wei, S.W., On vanishing theorems for vector bundle valued p-forms and their applications. Commun. Math. Phys. **304**, 329–368 (2011)
27. Dysthe, K.B.: Modelling a "rogue wave" – speculations or a realistic possibility? In: Olagnon, M., Athanassoulis, G. (eds.) Rogue Waves 2000, pp. 255–264. Ifremer, Brest (2001)
28. Edelen, D.G.B.: Applied Exterior Calculus. Wiley, New York (1985)
29. Ern, A., Guermond, J-L., Discontinnuous Galerkin methods for Friedrichs' systems. I. General theory, SIAM J. Numer. Anal. **44**, 753–778 (2006)
30. Ern, A., Guermond, J-L., Caplain, G.: An intrinsic criterion for the bijectivity of Hilbert operators related to Friedrichs' systems. Commun. Part. Differ. Equat. **32**, 317–341 (2007)

31. Felsen, L.B.: Evanescent waves. J. Opt. Soc. Amer. **66**, 751–760 (1976)
32. Felsen, L.B.: Novel ways for tracking rays, J. Optical Soc. Amer. A 2, 954–963 (1985)
33. Gilbarg, D., Trudinger, N.S.: Elliptic Partial Differential Equations of Second Order. Springer, Berlin (1983)
34. Gu, C.: Boundary value problems for quasilinear symmetric systems and their applications in mixed equations [in Chinese]. Acta Math. Sin. **21**, 119–129 (1978)
35. Gu, C.: On partial differential equations of mixed type in n independent variables. Commun. Pure Appl. Math. **34**, 333–345 (1981)
36. Gutshabashi, Ye. Sh., Lavrenov, I.V.: Swell transformation in the Cape Agulhas current. Izv. Atmos. Ocean. Phys. **22**, 494–497 (1986)
37. Heller, E.J., Kaplan, L., Dahlen, A.: Refraction of a Gaussian seaway. J. Geophys. Res. **113**, C09023, 14 (2008)
38. Irvine, D.E., Tilley, D.G.: Ocean wave directional spectra and wave-current interaction in the Alguhas from the shuttle imaging radar-B synthetic aperture radar. J. Geophys. Res. **93**, 15,389–15,401 (1988)
39. Isobe, T.: A regularity result for a class of degenerate Yang–Mills connections in critical dimensions. Forum Math. **20**, 1109–1139 (2008)
40. Janhunen, P.: Magnetically dominated plasma models of ball lightning. Annales Geophysicae. Atmos. Hydrospheres Space Sci. **9**, 377–380 (1991)
41. Jensen, M.: Discontinuous Galerkin methods for Friedrichs systems with irregular solutions, Ph.D. thesis, University of Oxford (2004)
42. Kambe, T.: Gauge principle for flows of an ideal fluid, Fluid Dyn. Res. **32**, 193–199 (2003)
43. Kapilevich, M.B.: On an equation of mixed elliptic-hyperbolic type [in Russian], Mat. Sbornik **30**, 11–38 (1952)
44. Karatoprakliev, G.D.: Equations of mixed type and degenerate hyperbolic equations in multidimensional domains [in Russian], Differencial'nye Uravn. **8**, 55–67 (1972)
45. Karatoprakliev, G.D.: A certain equation of mixed type in multidimensional domains [in Russian]. Dokl. Akad. Nauk SSSR **208**, 528–530 (1973) [Soviet Math. Dokl. **14**, 116–119 (1973)]
46. Karatoprakliev, G.D.: A class of mixed type equations in multidimensional domains [in Russian]. Dokl. Akad. Nauk SSSR **230**, 769–772 (1976) [Soviet Math. Dokl. **17**, 1379–1383 (1976)]
47. Karatoprakliev, G.D.: The formulation and solvability of boundary-value problems for equations of mixed type in multidimensional domains [in Russian]. Dokl. Akad. Nauk SSSR **239**, 257–260 (1978) [Soviet Math. Dokl. **19**, 304–308 (1978)]
48. Karatoprakliev, G.D.: Boundary-value problems for equations of mixed type in multidimensional domains [in Russian]. Partial Differential Equations, Warsaw, 1978, vol. 10, pp. 261–269. Banach Center Publications, Warsaw (1983)
49. Keller, J.B.: Surface waves on water of nonuniform depth. J. Fluid Mech. **4**, 607–614 (1958)
50. Khuri, M.A.: Boundary value problems for mixed type equations and applications. J. Nonlinear Anal. Ser. A: TMA (to appear); arXiv:1106.4000v1
51. Kuz'min, A.G.: Boundary value problems for equations of mixed type containing a mixed derivative [in Russian], Differentsial'nye Uravn. **22**, 66–74 (1986)
52. Ladyzhenskaya, O.A., Ural'tseva, N.N.: Linear and Quasilinear Elliptic Equations. Academic Press, New York (1968)
53. Lavrenov, I.: The wave energy concentration at the Agulhas current of South Africa. Nat. Hazards **17**, 117–127 (1998)
54. Lavrent'ev, M.A., Bitsadze, A.V.: On the problem of equations of mixed type [in Russian], Doklady Akad. Nauk SSSR (n.s.) **70**, 373–376 (1950)
55. Lerner, M.E., Repin, O.A.: A boundary value problem for an equation of mixed type on domains with multiconnected subdomains of hyperbolicity [in Russian]. Sibirskii Mat. Zh. **44**, 160–177 (2003)
56. Lobo, F., Crawford, P.: Time, closed timelike curves, and causality. In: Bucheri, R., Saniga, M., Stuckey, W.B. (eds.) The Nature of Time: Geometry, Physics, and Perception. NATO Science Series II: Mathematics, Physics, and Chemistry, Kluwer, Boston (2003)

57. Lupo, D., Micheletti, A.M., Payne, K.R.: Existence of eigenvalues for reflected Tricomi operators and applications to multiplicity of solutions for sublinear and asymptotically linear nonlical Tricomi problems. Adv. Differ. Equat. **4**, 391–412 (1999)
58. Lupo, D., Morawetz, C.S., Payne, K.R.: On closed boundary value problems for equations of mixed elliptic-hyperbolic type. Commun. Pure Appl. Math. **60**, 1319–1348 (2007)
59. Lupo, D., Payne, K.R.: Existence of a principal eignevalue for the Tricomi problem. Proceedings of the Conference on Nonlinear Differential Equations (Coral Gables, Fl., 1999), pp. 173–180, Electronic J. Differential Equations Conf. **5**, Southwest Texas State University, San Marcos, TX (2000)
60. Lupo, D., Payne, K.R.: A dual variational approach to a class of nonlocal semilinear Tricomi problems. Nonlinear Differ. Equat. Appl. **6**, 247–266 (1999)
61. Lupo, D., Payne, K.R.: The dual variational method in nonlocal semilinear Tricomi problems. Nonlinear Analysis and its Applications to Differential Equations (Lisbon, 1998) Progr. Nonlinear Diff Equations Appl., vol. 43, pp. 321–338. Birkhäuser Boston, Boston (2001)
62. Lupo, D., Payne, K.R.: On the maximum principle for generalized solutions to the Tricomi problem. Commun. Contemp. Math. **2**, 535–557 (2000)
63. Lupo, D., Payne, K.R.: Spectral bounds Tricomi problems and application to semilear existence and existence with uniqueness results. J. Differ. Equat. **184**, 139–162 (2002)
64. Lupo, D., Payne, K.R.: Critical exponents for semilinear equations of mixed elliptic-hyperbolic and degenerate types. Commun. Pure Appl. Math. **56**, 403–424 (2003)
65. Lupo, D., Payne, K.R., Popivanov, N.I.: Nonexistence of nontrivial solutions for supercritical equations of mixed elliptic-hyperbolic type. Contributions to nonlinear analysis, Progress in Nonlinear Differential Equations Applications, vol 66, pp. 371–390. Birkhäuser, Basel (2006)
66. Magnanini, R., Talenti, G.: On complex-valued solutions to a 2D eikonal equation. Part one: qualitative properties. Contemp. Math. **283**, 203–229 (1999)
67. Magnanini, R., Talenti, G.: Approaching a partial differential equation of mixed elliptic-hyperbolic type. In: Anikonov, Yu.E., Bukhageim, A.L., Kabanikhin, S.I., Romanov, V.G. (eds.) Ill-posed and Inverse Problems, pp. 263–276. VSP, Utrecht (2002)
68. Magnanini, R., Talenti, G.: On complex-valued solutions to a two-dimensional eikonal equation. II. Existence theorems. SIAM J. Math. Anal. **34**, 805–835 (2003)
69. Magnanini, R., Talenti, G.: On complex-valued solutions to a 2D eikonal equation. III. Analysis of a Bäcklund transformation. Appl. Anal. **85**, 249–276 (2006)
70. Magnanini, R., Talenti, G.: On complex-valued 2D eikonals. IV. Continuation past a caustic. Milan J. Math. **77**, 1–66 (2009)
71. Marathe, K.B., Martucci, G.: The Mathematical Foundations of Gauge Theories. North–Holland, Amsterdam (1992)
72. Marini, A., Otway, T.H.: Nonlinear Hodge-Frobenius equations and the Hodge-Bäcklund transformation. Proc. R. Soc. Edinburgh, Ser. A **140**, 787–819 (2010)
73. Mei, C.C.: The Applied Dynamics of Ocean Surface Waves. World Scientific, Singapore (1989)
74. Mei, C.C., Tlapa, G.A., Eagelson, P.S.: An asymptotic theory for water waves on beaches of mild slope J. Geophys. Res. **73**, 4555–4560 (1968)
75. Xu, M., Yang, X-P.: Existence of distributional solutions of closed Dirich-let problem for an elliptic-hyperbolic equation. J. Nonlinear Analysis Ser. A: TMA, to appear
76. Morawetz, C.S.: A weak solution for a system of equations of elliptic-hyperbolic type. Commun. Pure Appl. Math. **11**, 315–331 (1958)
77. Morawetz, C.S., Stevens, D.C., Weitzner, H.: A numerical experiment on a second-order partial differential equation of mixed type. Commun. Pure Appl. Math. **44**, 1091–1106 (1991)
78. Morrey, C.B.: Multiple Integrals in the Calculus of Variations. Springer, Berlin (1966)
79. Müller-Rettkowski, A.H.: Existence of a characteristic boundary value problem for an equation of mixed type in three-dimensional space. Math. Methods Appl. Sci. **5**, 346–355 (1983)

80. Otway, T.H.: An asymptotic condition for variational points of nonquadratic functionals. Ann. Fac. Sci. Toulouse **9** 187–195 (1990)
81. Otway, T.H.: The coupled Yang-Mills-Dirac equations for differential forms. Pacific J. Math. **146**, 103–113 (1990)
82. Otway, T.H.: Properties of nonlinear Hodge fields. J. Geom. Phys. **27**, 65–78 (1998)
83. Otway, T.H.: Nonlinear Hodge maps. J. Math. Phys. **41**, 5745–5766 (2000)
84. Otway, T.H.: Variational equations on mixed Riemannian-Lorentzian metrics. J. Geom. Phys. **58**, 1043–1061 (2008)
85. Otway, T.H.: Unique solutions to boundary value problems in the cold plasma model. SIAM J. Math. Anal. **42**, 3045–3053 (2010)
86. Paskalev, G.P.: Nonlocal boundary value problem for a class of high order partial differential equations. C. R. Acad. Bulgare Sci. **53**, 13–16 (2000)
87. Paskalev, G.P.: A nonlocal boundary value problem for a higher-order mixed equation [in Russian]. Differentsial'nye Uravn. **36**, 393–399 (2000) [Differential Equations **36**, 441–448 (2000)]
88. Paskalev, G.P.: Sufficient conditions for the smoothness of the generalized solution of a nonlocal boundary value problem for a higher–order mixed–type equation [in Russian]. Differentsial'nye Uravn. **36**, 799–805 (2000) [Differential Equations **36**, 886–893 (2000)]
89. Payne, K.R.: Propagation of singularities for solutions to the Dirichlet problem for equations of Tricomi type, Rend. Sem. Mat. Univ. Pol. Torino **54**, 115–137 (1996)
90. Payne, K.R.: Solvability theorems for linear equations of Tricomi type. J. Mat. Anal. Appl. **215**, 262–273 (1997)
91. Payne, K.R.: Singular metrics and associated conformal groups underlying differential operators of mixed and degenerate types. Ann. Mat. Pura Appl. **184**, 613–625 (2006)
92. Peregrine, D.H.: Interaction of water waves and currents. Adv. Appl. Mech. **16**, 9–117 (1976)
93. Pierson, Jr., W.J.: The interpretation of crossed orthogonals in wave refraction phenomena. Tech. Memo. vol. 21, Beach Erosion Board (1951)
94. Pilant, M.: The Neumann problem for an equation of Lavrent'ev-Bitsadze type. J. Math. Anal. Appl. **106**, 321–359 (1985)
95. Piliya, A.D., Fedorov, V.I.: Singularities of the field of an electromagnetic wave in a cold anisotropic plasma with two-dimensional inhomogeneity. Sov. Phys. JETP **33**, 210–215 (1971)
96. Popivanov, N.I.: A certain boundary value problem for an equation of mixed type with two perpendicular curves of parabolic degeneracy [in Russian]. C. R. Acad. Bulgare Sci. **25**, 441–444 (1972)
97. Popivanov, N.I.: Equations of mixed type with two lines of degeneracy in unbounded domains. I. Imbedding theorem. Uniqueness theorem [in Russian]. Differentsial'nye Uravnen. **14**, 304–317 (1978) [Differential Equations **14**, 212–221 (1978)]
98. Popivanov, N.I.: Equations of mixed type with two lines of degeneracy in unbounded domains. II. Existence of a strong solution [in Russian]. Differentsial'nye Uravnen. **14**, 665–679 (1978) [Differential Equations **14**, 468–479 (1978)]
99. Popivanov, N.I.: The theory of linear systems of first order partial differential equations, and equations of mixed type with two perpendicular lines of parabolic degeneracy. In: Mathematics and Mathematical Education. Proc. Second Spring Conf. Bulgarian Math. Soc., Vidin, 1973, [in Bulgarian], pp. 175–181. Izdat. Bulgar. Akad. Nauk., Sofia (1974)
100. Popivanov, N.I., Schneider, M.: Nonlocal regularization of Protter problem for the 3-D Tricomi equation. Proceedings of the Eighth International Colloquium on Differential Equations, Plovdiv, 1997, pp. 373–378. VSP, Utrecht (1998)
101. Protter, M.H.: New boundary value problem for the wave equation and equations of mixed type. J. Rat. Mech. Anal. **3**, 435–446 (1954)
102. Rassias, J.M.: On the Tricomi problem with two parabolic lines of degeneracy. Bull. Inst. Math. Acad. Sin. **12**, 51–56 (1984)
103. Rassias, J.M.: The exterior Tricomi and Frankl problems for quaterelliptic-quaterhyperbolic equations with eight parabolic lines. Eur. J. Pure Appl. Math. **4**, 186–208 (2011)

104. Shcherbakov, E.A.: The electrostatic Green's function for the Tricomi equation and home-omorphic solutions of a degenerate elliptic system associated with it. Russian Mathematics (Izvestiya VUZ Matematika), **35**, 60–66 (1991)
105. Shen, M.C., Meyer, R.E., Keller, J.B.: Spectra of water waves in channels and around islands. Phys. Fluids **11**, 2289–2304 (1968)
106. Sibner, L.M.: A boundary value problem for an equation of mixed type having two transitions. J. Differ. Equat. **4**, 634–645 (1968)
107. Sibner, L.M., Sibner, R.J., Yang, Y.: Generalized Bernstein property and gravitational strings in Born–Infeld theory. Nonlinearity **20**, 1193–1213 (2007)
108. Smith, R.: Giant waves. J. Fluid Mech. **77**, 417–431 (1976)
109. Smoller, J.: Shock Waves and Reaction-Diffusion Equations. Springer, Berlin (1983)
110. Smoller, J., Temple, B.: Astrophysical shock wave solutions of the Einstein equations. Phys. Rev. D **51**, 2733–2743 (1995)
111. Sorokina, N.G.: Strong solvability of the Tricomi problem [in Russian]. Ukrain. Mat. Zh. **18**, 65–77 (1966)
112. Sorokina, N.G.: Strong solvability of the generalized Tricomi problem [in Russian]. Ukrain. Mat. Zh. **24**, 558–561 (1972) [Ukrainian Math. J. **24**, 451–453 (1973)]
113. Sorokina, N.G.: Strong solvability of a boundary value problem for an equation of mixed type in multidimensional domains [in Russian]. Ukrain. Mat. Zh. **26**, 115–123 (1974)
114. Sorokina, N.G.: Two boundary value problems for multidimensional equations of mixed type that are degenerate on a circular cylinder [in Russian]. Ukrain. Mat. Zh. **27**, 614–623 (1975)
115. Sorokina, N.G.: Multidimensional generalizations of C. Morawetz' theorem on the weak solvability of a system of mixed type differential equations [in Russian]. Differentsial'nye Uravn. **13**, 155–158 (1977)
116. Sorokina, N.G.: Boundary value problems for a three-dimensional equation of mixed type of the second kind [in Russian]. Differentsial'nye Uravn. **18**, 173–176 (1982)
117. Stewart, J.M.: Signature change, mixed problems and numerical relativity. Class. Quantum Grav. **18**, 4983–4995 (2001)
118. van Stockum, W.J.: The gravitational field of a distribution of particles rotating about an axis of symmetry. Proc. R. Soc. Edinburgh **57**, 135–154 (1937)
119. Tipler, F.J.: Rotating cylinders and the possibility of causality violation. Phys. Rev. D **9**, 2203–2206 (1974)
120. Toffoli, A., Ardhuin, F., Babanin, A., Benoit, M., Bitner–Gregerseb, E.N., Cavaleri, L., Monbaliu, J., Onorato, M., Osborne, A.R.: Extreme waves in directional wave fields travers-ing uniform currents. Proceedings of the HYDRALAB III Joint User Meeting, Hannover, February 2010 http://www.hydraloab.eu/hydralabIII/proceeding_ta_projects.asp. Cited 2 Aug 2011
121. Torre, C.G.: The helically reduced wave equation as a symmetric positive system. J. Math. Phys. **44**, 6223–6232 (2003)
122. Torre, C.G.: Uniqueness of solutions to the helically reduced wave equation with Sommerfeld boundary conditions. J. Math. Phys. **47**, 073501 (2006)
123. Tricker, R.A.R.: Bores, Breakers, Waves, and Wakes. American Elsevier, New York (1965)
124. Tso, K. (Chou, K-S.): Nonlinear symmetric positive systems. Ann. Inst. Henri Poincaré **9**, 339–366 (1992)
125. Tsui, K.H.: A self-similar magnetohydrodynamic model for ball lightnings. Phys. of Plasmas **13**, 072102 (2006)
126. Vakhania, N.: On boundary value problems for the hyperbolic case, J. Complex. **10**, 341–355 (1994)
127. Ward, S.N., Asphaug, E., Asteroid impact tsunami of 2880 March 16. Geophys. J. Int. **153**, F6–F10 (2003)
128. White, B.S., Fornberg, B.: On the chance of freak waves at sea. J. Fluid Mech. **355**, 113–138 (1998)

129. Wu, S.: Well-posedness in Sobolev spaces of the full water wave problem in 3-D. J. Am. Math. Soc. **12**, 445–495 (1999)
130. Xu, Z-F., Xu, Z-Y., Chen, G-L.: Some notes on the Busemann equation [in Chinese]. Adv. in Math. (Beijing) **16**, 81–86 (1987)
131. Yagdjian, K.: A note on the fundamental solution for the Tricomi-type equation in the hyperbolic domain. J. Diff. Eq. **206**, 227–252 (2004)
132. Zakharov, V.E.: Stability of periodic waves of finite amplitude in the surface of a deep fluid. Sov. Phys. J. Appl. Mech. Tech. Phys. **4**, 190–194 (1968)
133. Zheng, Y.: Systems of Conservation Laws: Two-Dimensional Riemann Problems. Birkhauser, Boston (2001)
134. Zheng, Y.: A global solution to a two-dimensional Riemann problem involving shocks as free boundaries. Acta Math. Applicatae Sin. **19**, 559–572 (2003)
135. Zheng, Y.: Weakly nonlinear asymptotic models of the multi–D compressible Euler equations, e-print. http://www.math.psu/yzheng/PG/sjtu-cnusli.ps. Cited 2 Aug 2011

Index

LECTURE NOTES IN MATHEMATICS

Edited by J.-M. Morel, B. Teissier; P.K. Maini

Editorial Policy (for the publication of monographs)

1. Lecture Notes aim to report new developments in all areas of mathematics and their applications - quickly, informally and at a high level. Mathematical texts analysing new developments in modelling and numerical simulation are welcome.

   Monograph manuscripts should be reasonably self-contained and rounded off. Thus they may, and often will, present not only results of the author but also related work by other people. They may be based on specialised lecture courses. Furthermore, the manuscripts should provide sufficient motivation, examples and applications. This clearly distinguishes Lecture Notes from journal articles or technical reports which normally are very concise. Articles intended for a journal but too long to be accepted by most journals, usually do not have this "lecture notes" character. For similar reasons it is unusual for doctoral theses to be accepted for the Lecture Notes series, though habilitation theses may be appropriate.

2. Manuscripts should be submitted either online at www.editorialmanager.com/lnm to Springer's mathematics editorial in Heidelberg, or to one of the series editors. In general, manuscripts will be sent out to 2 external referees for evaluation. If a decision cannot yet be reached on the basis of the first 2 reports, further referees may be contacted: The author will be informed of this. A final decision to publish can be made only on the basis of the complete manuscript, however a refereeing process leading to a preliminary decision can be based on a pre-final or incomplete manuscript. The strict minimum amount of material that will be considered should include a detailed outline describing the planned contents of each chapter, a bibliography and several sample chapters.

   Authors should be aware that incomplete or insufficiently close to final manuscripts almost always result in longer refereeing times and nevertheless unclear referees' recommendations, making further refereeing of a final draft necessary.

   Authors should also be aware that parallel submission of their manuscript to another publisher while under consideration for LNM will in general lead to immediate rejection.

3. Manuscripts should in general be submitted in English. Final manuscripts should contain at least 100 pages of mathematical text and should always include

   – a table of contents;
   – an informative introduction, with adequate motivation and perhaps some historical remarks: it should be accessible to a reader not intimately familiar with the topic treated;
   – a subject index: as a rule this is genuinely helpful for the reader.

   For evaluation purposes, manuscripts may be submitted in print or electronic form (print form is still preferred by most referees), in the latter case preferably as pdf- or zipped psfiles. Lecture Notes volumes are, as a rule, printed digitally from the authors' files. To ensure best results, authors are asked to use the LaTeX2e style files available from Springer's web-server at:

   ftp://ftp.springer.de/pub/tex/latex/svmonot1/ (for monographs) and
   ftp://ftp.springer.de/pub/tex/latex/svmultt1/ (for summer schools/tutorials).

Additional technical instructions, if necessary, are available on request from lnm@springer.com.

4. Careful preparation of the manuscripts will help keep production time short besides ensuring satisfactory appearance of the finished book in print and online. After acceptance of the manuscript authors will be asked to prepare the final LaTeX source files and also the corresponding dvi-, pdf- or zipped ps-file. The LaTeX source files are essential for producing the full-text online version of the book (see http://www.springerlink.com/openurl.asp?genre=journal&issn=0075-8434 for the existing online volumes of LNM). The actual production of a Lecture Notes volume takes approximately 12 weeks.

5. Authors receive a total of 50 free copies of their volume, but no royalties. They are entitled to a discount of 33.3 % on the price of Springer books purchased for their personal use, if ordering directly from Springer.

6. Commitment to publish is made by letter of intent rather than by signing a formal contract. Springer-Verlag secures the copyright for each volume. Authors are free to reuse material contained in their LNM volumes in later publications: a brief written (or e-mail) request for formal permission is sufficient.

Addresses:
Professor J.-M. Morel, CMLA,
École Normale Supérieure de Cachan,
61 Avenue du Président Wilson, 94235 Cachan Cedex, France
E-mail: morel@cmla.ens-cachan.fr

Professor B. Teissier, Institut Mathématique de Jussieu,
UMR 7586 du CNRS, Équipe "Géométrie et Dynamique",
175 rue du Chevaleret
75013 Paris, France
E-mail: teissier@math.jussieu.fr

For the "Mathematical Biosciences Subseries" of LNM:

Professor P. K. Maini, Center for Mathematical Biology,
Mathematical Institute, 24-29 St Giles,
Oxford OX1 3LP, UK
E-mail : maini@maths.ox.ac.uk

Springer, Mathematics Editorial, Tiergartenstr. 17,
69121 Heidelberg, Germany,
Tel.: +49 (6221) 4876-8259

Fax: +49 (6221) 4876-8259
E-mail: lnm@springer.com